Dynamite Stories

T R A N S M O N T A N U S   1 1

Published by New Star Books
Series Editor: Terry Glavin

Other books in the Transmontanus series

Dynamite Stories

Judith Williams

TRANSMONTANUS / NEW STAR BOOKS VANCOUVER

Norman and Doris Hope with their dog Stinky at Refuge Cove, late 1940s.

To the Duke and Duchess of West Redonda.

'Economists think that economic indicators are the metaphors for humanity. Novelists think that stories are the true indicators of human existence. [Bruce] Chatwin correctly saw stories as paradigms of humanity.'

NICHOLAS SHAKESPEARE

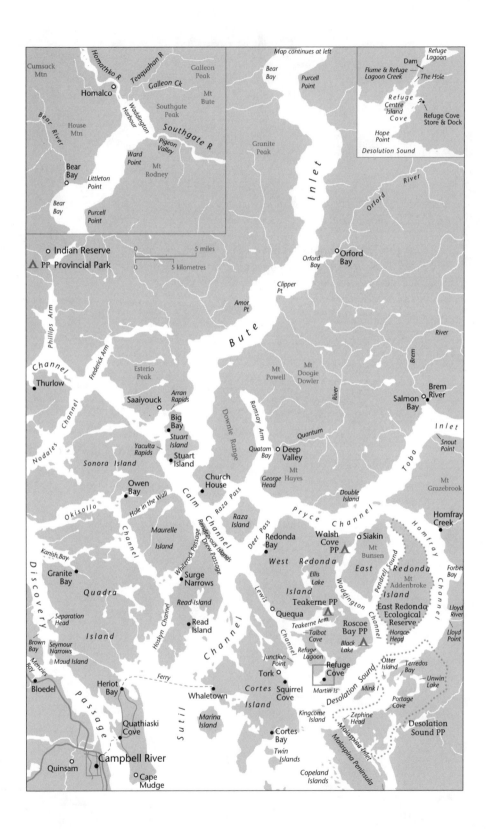

Inset (top left):

Cumsack Mtn

Homathko R

Teaquahon R

Galleon Peak

Galleon Ck

Homalco

Mt Bute

Southgate Peak

Waddington Harbour

Southgate R

Bear River

House Mtn

Pigeon Valley

Ward Point

Mt Rodney

Bear Bay

Littleton Point

Bear Bay

Purcell Point

Legend:

o Indian Reserve

▲ PP Provincial Park

0   5 miles

0   5 kilometres

Inset (top right):

Refuge Lagoon

Dam

Flume & Refuge Lagoon Creek

The Hole

Refuge Centre Island Cove

Refuge Cove Store & Dock

Hope Point

Desolation Sound

Main map:

Map continues at left

Bear Bay

Purcell Point

Inlet

Orford River

Granite Peak

Orford Bay

Orford Bay

River

Clipper Pt

Phillips Arm

Amor Pt

Brem

Channel

Frederick Arm

Esterio Peak

Bute

Mt Powell

Mt Doogie Dowler

River

Brem River

Thurlow

Salmon Bay

Inlet

Nodales Channel

Arran Rapids

Saaiyouck

Downie Range

Ramsay Arm

Quantum

Snout Point

Big Bay

Quatam Bay

Deep Valley

Stuart Island

Yaculta Rapids

Stuart Island

Sonora Island

Mt Hayes

Toba

Mt Grazebrook

George Head

Double Island

Owen Bay

Hole in the Wall

Calm Channel

Raza Pass

Pryce Channel

Homfray Creek

Okisollo

Raza Island

Deer Pass

Redonda Bay

Walsh Cove PP

Siakin

Forbes Bay

Channel

Maurelle Island

Whiterock Passage

Rendezvous Islands

Drew Passage

West Redonda

Mt Bunsen

East Redonda

Pendrell Sound

Homfray Channel

Kanish Bay

Island

Ellis Lake

Mt Addenbroke

Granite Bay

Surge Narrows

Lewis Channel

Teakerne PP

East Redonda Ecological Reserve

Lloyd River

Discovery

Quadra

Read Island

Quequa

Waddington Channel

Horace Head

Lloyd Point

Separation Head

Read Island

Teakerne Arm

Roscoe Bay PP

Island

Hoskyn Channel

Talbot Cove

Black Lake

Brown Bay

Seymour Narrows

Junction Point

Refuge Lagoon

Otter Island

Terredos Bay

Menzies Bay

Maud Island

Ferry

Tork

Refuge Cove

Unwin Lake

Bloedel

Cortes

Squirrel Cove

Martin Is

Mink

Heriot Bay

Whaletown

Island

Kingcome Island

Portage Cove

Passage

Marina Island

Cortes Bay

Zephine Head

Desolation Sound PP

Quathiaski Cove

Sutil

Malaspina Inlet

Malaspina Peninsula

Campbell River

Twin Islands

Quinsam

Cape Mudge

Copeland Islands

Contents

Robert Donley's Refuge Cove Store circa 1916. COURTESY THE
COLLECTION OF RITA HIGGS

Prologue

In 1918 Robert Donley established the Donley Trading Company on West Redonda Island, in Desolation Sound, one hundred miles north of Vancouver. Donley's Refuge Cove store stocked groceries and hardware and his tippy log dock served as a fish-buying station and a depot for kerosene, gas and oil. A post office was added, and by 1925 the Refuge Cove Store sign stood at light point.

Doris and Norman Hope first appeared on the Desolation Sound stage after disembarking from the Union Steamship the *Cardero*, on Saint Patrick's Day, 1945. With Norman's brother Buster, and his wife Vivian, they purchased the buildings, docks and 186 acres of land from Jack Tindall, who had owned it for 15 years.

When 18 shareholders formed The Refuge Cove Land and Housing Co-op on the Hopes' acres in 1972, we did not so much buy a piece of land as invest our futures in a tangled web of west coast history. Fortunately, Norman and Doris spent the rest of their lives *in situ* sorting out all the connections for us. After Norman passed on, Doris continued to live in the old floathouse jacked up on the boardwalk behind the store. There she entertained and instructed the people who occasion these stories.

Ken MacPherson and Gordie Irwin working the bundler.

If two sticks are good, four are better

The Tales of the Redonda Woods began their transformation from oral to textual form on the evening of August 19, 1994. The Refugees had gathered for dinner on what John Dixon liked to style his "pier." Like our other Redonda neighbour Barry Ketcheson, John had extended the log pilings and decking from shore to sea. We thought that in time he might float a log raft somewhere in front of the pier just as Barry had done but would he, like Barry, neglect to connect the two? When a plane crashed near Barry's place, the investigating Coast Guard stood gazing at the lacuna, the void where a ramp to shore should have been. Still isn't. As usual, Desolation Sound project completion was skewed by an individual's hidden agenda.

Dinner was a 32-pound salmon that had been fresh-smoked in honour of Refuge doyenne Doris Hope who, citing ill health or, depending on who told the story, the presence of dogs and children, had at the last moment declined to attend. Her nephew-in-law Rick Carter asked us what we had planned for Aunty Doris' 50th Refuge anniversary. Embarrassingly, the answer was "nothing," and I wondered why. Doris, although 83, always seemed current. We were together in time, and I began to consider how her address to the world caused me to perceive her that way. She certainly talked about the past. After I'd visited some empty nearby bay or channel it was my habit to call in and be debriefed.

"Doris, I just got back from Homfray Creek."

"Did you dear," she would say and light a cigarette. "One hundred and eighty men." She'd pause, put her foot up on a stool, and settle in. "One hundred and eighty men worked at Homfray in 1945."

I'd just seen a pristine creek fall heart-achingly through sandstone pools, green dreams and five-foot ferns into a placid channel. I tend to date habitation, Native or white, in terms of shell depth, rusted iron and/or era of broken china laid out on a beach. I have a major collection of blue patterned fragments: bagged, sorted and labelled. There was no sign of habitation at Homfray Creek, not a rusted nail.

Doris would wave her cigarette barward saying, "If that husband of yours would fix me a drink I'll go on." This meant I should stow my fragment theory so she could get on with the true story: her story.

"One hundred and eighty men worked there logging. Dorothy and Ed Thomas operated the camp boat Devon that took men from Refuge to Homfray Creek where a boatman would pick them up and transport them to the bunkhouses."

"How'd they get to Refuge?"

"On the Union Steamships of course. Twice a week. "One day, I looked out and saw a young woman on the float getting into the Devon. She was all done up: dress, high heeled shoes my dear, stockings, hat, the lot. When she got to Homfray the boatman came to pick her up — Oh! those old boats! All haywired together. You don't know! She got in and the boatman . . ."

"Drunk?" I ventured.

"Of course! He revved up the engine and the flywheel stripped the dress right off her. Later the bastard sat right here, saying, 'God damn it. If I'd only revved it up a little higher. . . .' 'You wouldn't?' I said. 'I would,' he said. 'I could have got the lot off her.' Hee, hee, hee."

Life, as Doris tells it, is a narrative of past events becoming present in the telling. Oral history evolves, transforms itself and arrives at the dock, in a dugout canoe with an outboard engine, looking for a cappuccino. A great storyteller, and Doris was certainly that, tells the same story over and over, reworking its structure and emphasis to suit the audience and era.

It was interesting to listen to Doris and her pal Nellie Black dispute local dates and facts: who married whom and who lived where, which is often confusing since upcoast dwellings were themselves peripatetic. Before the Co-op bought the Hopes' 186 acres no one had built a land house from scratch in the Cove itself. All houses were either afloat or had been floathouses that were winched up onto land. It was considered radical, not to say shortsighted, to actually build houses that would not, or could not, be made to float away.

Doris urged us to abandon plans to build frame or log houses in favour of trailers, and up and down the coast that is presently the accepted way of setting up camp. Every so often, I see a stark white, plastic-sided box being towed north on a barge. Will it be as charming collapsing seaward as did our "Middle Cabin," the only trailer we ever allowed on our land. It was so traumatized by the move that it leaked like the hose that was chewed by the bear. Mushrooms thrived in its innards until we gleefully burned it into to melted pools of aluminum and slowly revolving, flaming tires.

When Norman and Doris joined the Refuge Cove community in 1945 all the houses were on log floats and strung in rows along boom sticks with cable and chain. Walking a boom stick was how you got from A to B — man, woman, child or dog. According to Doris, one of the residents, Birdie MacPherson, was a large woman who kept a neat floathouse and had at that time a two-year-old daughter. Her husband Ken, ultimately one of the last one-man logging shows around, was as lean and taciturn as a leather bootlace. The MacPhersons were rafted up next to "the Norwegian," in line with what Doris always referred to as "the Bachelors" in tones that intimated they were an exotic tribe in need of constant, though ultimately futile, correction. I never quite understood why until we grew some of our own.

There was only one problem with a floathouse, only one reason to jack it up on land: teredos. These small, perhaps likable, even ecologically important worms burrow into cedar float logs and reduce them to calcified tunnels. Teredos can become endemic to a logging area, munch up the profits, and drive a camp out of a bay. The received wisdom is that the way to eliminate teredos is with dynamite. The shock, it is claimed, lays them out like mummies in their burrows.

Ken MacPherson woke early one quiet Sunday morning with a kind of listing feeling. The floor was not quite level. Deciding to nip that in the bud, or in this case worm, he went out, attached sticks of dynamite to the float and, remembering The First Law of Dynamite: If two sticks are good, four are better, he added a couple of extra sticks to the front floater and lit the fuse.

It worked. The dynamite exploded, the teredos folded their hands, if they had any, over their breasts and snuffed it. Birdie exited the floathouse with a frying pan in her hand.

"MacPherson, if I catch ya you'll be even shorter than ya are now!" or, depending on who told this story, Norman or Doris, "You little pricksucker I'll have the balls off ya!" Uninformed of the impending teredo shock wave, Birdie had been frying up a mess of sausages for breakfast. The explosion flung the pan from the stove, wound the string of sausages round the lamp, broke all the dishes and blew the kiddy off the potty and across the room.

The neighbours, roused by the blast, turned out to watch the fight and then repaired to the floathouse of the Norwegian to soothe their nerves with a touch of his home-brew, known as the Blessed Sacrament. A visitor who'd lived for many years at Walsh Cove on the northeast side of West Redonda Island joined the party. Eventually, feeling in a sufficient state of grace, he decided to go home and as his fish boat *fu-chug . . . fu-chug . . . fu-chug*ged by, the brewmaster, in a excess of *bonhomie*, threw a bottle of the Blessed Sacrament onto his deck. The man from Walsh Cove turned left at the Refuge light and headed around the south tip of Redonda and home. Still celebrating MacPherson's triumph over the teredos, the assembled were surprised when some hours later the man appeared from the north, having by necessity circumnavigated West Redonda. Rejoining the gathering, he claimed he hadn't been able to find Walsh Cove. A chart was produced, and it was pointed out to him that to circle the island, one must inevitably pass that cove.

MacPherson's cabin in "the Hole", formerly the Black family's house. PHOTO BY JUDITH WILLIAMS, 1988

By the time the Co-op was formed, Birdie had removed herself to a house situated on *terra firma* and bought Ken a fast boat in which to commute. He camped out in Grandpa

Black's old floathouse, which still bears tattooed evidence, in the green, ferny linoleum, of caulk boots worn to dinner. MacPherson, hipless, hung his Bannockburn pants high on his braces, wore his hardhat at a steep angle and kept a mickey of Maalox in his back pocket to soothe a stomach that'd had too many close calls logging.

In the late 1970s Ken undertook to move Paul Emmons' floathouse up onto our land. It was inched down from its perch on the Black family's small bite out of the Co-op's acres and onto a float for the night. On the next day's high tide it was to be floated to shore and winched up onto the end of Co-op property beside Lagoon Creek.

Ken needed the Maalox the second day of the house move. The floating, log bundler of rusted wire and gears broke down mid-winch, and the house listed forward at an alarming angle. The tide ebbed. Tense repairs had to be made. We hung around embarrassed. We hoped it would be okay, and wished them well, but we no one was about to miss any action, good or bad. I was pooping around in a row boat taking what I claimed were important documentary photographs. Suddenly the pile of nuts and bolts came to life. The Maalox was consulted, "second mate" Gordie Irwin lifted his gold leafed hardhat, wiped his brow, and the house waddled up the beach to the cheers of the flotilla of onlookers. The Gilchrist jacks were inserted, the house raised above the high tide and props inserted. There it still stands. When the tide rises above the fifth stair, we know it's high.

Moving Paul's floathouse in the late 1970s. Pat Lovell is in the foreground and Glynne Evans is in the kayak.

PHOTO BY JUDITH WILLIAMS

I might not have begun to note down the Laws of Dynamite if sometime after hearing the teredo story from Doris, the Co-op hadn't decided that we should repair the boardwalk. It had been built to connect the floathouses that were, now in their old age, jacked up to roost along the steep shore from the store to the point. There wasn't really anywhere to put your foot, so the cabins had been joined by a tremulous series of planks laid

lengthwise from uprights jammed into available rock crevices. The splintered boards, punctured by thousands of holes caused by generations of caulk boots at dawn, sagged seaward in their middles. So alarming was their sag, so narrow were the boards, so high at low tide were you over the my-God-for-sure-this-time-the-board-will-give-way-and-I'm-doomed-rock, that when the Gibbons came to spend 1971 in the "End Cabin," Denise's nerve failed her. She begged Norm to go on ahead and take the children inside the house so they wouldn't see their mother crawl on hands and knees past the Hopes' house, past the "Middle Cabin," suspended — and certainly not for long — by a rusted cable from a rotting stump, to the "End Cabin."

It was well before Doris' 11:00 a.m. wake-up smoke, and any noise would occasion a window to be flung open and a stentorian voice announce, "All I ask is some simple consideration. People might have the good manners to keep the noise down so a person can get a little rest. Thank *you!*"

By 1973 the boardwalk was clearly unsafe, so the Co-op decided to replace it. Uncle Norman declared this was a moment he'd been waiting for. He'd blow "That pricksucker of a rock out of there" so the boardwalk would not slant seaward, nor take a sudden jump that hit your toe like a viper in the night.

Uncle Norman really did talk like that and occasionally, so do I, having been addressed by Norman in loggerise during countless poker games because of my shameful lack of card sense. This usually occurred after Norman had made a trip to the john and I, but no one else, had seen him tip a bottle of rye vertical and slug down several ounces. He'd swear, and Doris would remonstrate with, "Mind your language you bloody bastard."

As I was saying before I interrupted myself: we agreed to replace the boardwalk.

The scene: Uncle Norman assembling weapons with which to attack the offending rock.

The audience: an interested tourist with a large yacht tucked into the always highly desirable position behind the dock U and not all that far from the Hopes' house or the scene of action.

The cast: one self-defined old fart and five helpers, accused by the old fart of, "standing around useless as spare pricks at a wedding."

The tourist, sensing entertainment, poured himself a drink and settled into a deckchair on his flying bridge, which was conveniently level with the work party.

Uncle Norman, swearing a blue streak, drilled holes in the rock and set the dynamite.

People ran for cover.

Norman lit the fuse, ducked behind a outcrop, and waited.

The tourist waited. We all waited and, sure enough, MacPherson's Second Law of Dynamite held true: If you use enough you won't have to do it twice.

The 100-pound offending hunk of West Redonda parted from the matrix intact, went straight up in the air, arced to the left, and entered the water inches from where the tourist sat with his hand frozen around his glass. Fortunately or not, depending on who tells the story, the rock descended a generous two inches from the gunnels of his boat.

"That," Norman said, "will teach the asshole to leave his generator on all night."

Just what constituted the substance Uncle Norman used to part Redonda minor from Redonda major is useful to consider. Its essential compound seems to have been discovered through a series of events haywire enough to have the true Refuge Cove flavour. A dog, as usual, was involved.

The heart disease known as *angina pectoris* is characterized by intense pain. That nitroglycerine, the main ingredient of dynamite, blasting gelatine and gelignite, is used in its treatment was the result of the remarkably offhand researches of one Ascanio Sobrero. In 1847, for reasons that I still haven't gotten clear, Ascanio poured a concentrated mix of sulphuric and nitric acids into glycerine, thereby creating a yellowish oil that exploded and drove the container's glass fragments into his hands and face. Excited by the result, he licked off a drip and, finding it sweet, placed a small sample on his tongue which gave rise to, as he put it, "a most pulsating, violent headache accompanied by great weakness of the limbs." An undocumented weakness of the head allowed him to give a passing dog a few centigrams of what we now know as nitroglycerine. The poor animal instantly foamed at

the mouth. After a dose of ammonia and olive oil was adminis-
tered, the dog revived somewhat but "remained whining, trem-
bling violently and beating its head against the wall." This
sobered up Ascanio who ceased his experiments and announced,
and certainly in this he differs from many succeeding scientists,
"that science should not be made a pretext or means of . . . business speculations." No one paid any attention.

The first medical trials of nitroglycerine were conducted in
the field of homeopathy, which is based on the principal that
minute amounts of like cures like. Ah! What caused headaches
might cure them. (Uncle Norman, although not well remembered for of his homeopathic theories did, however, regularly pre-
scribe the hair of the dog as hangover cure.) When in 1858 Sir
Thomas Lauder Brunton discovered in 1858 that nitroglycerine
caused blood vessels to dilate, it became the effective treatment
for angina. Interestingly, a nitro patch can be attached to a suf-
ferer's body but, as with all dynamite transport, there is many a
slip between cup and lip. The nitro patch has been known to
explode with a sharp but non-lethal bang when the patient was
subjected to electroshock. Just why someone with heart disease,
and charged with dynamite, would be given electroshock is
another story.

There were other hazards. Workers processing the material for
explosives often suffered from "NG head" before becoming accli-
matized to it. When they took a vacation they lost their acclimati-
zation, and re-entry was rough. As a result they often took with
them, much against the rules, a small personal supply of the
explosive to sniff *en vacance*.

One of the great ironies of the history of dynamite is that it was
eventually brought sort of under control by the Swedish chemist,
Alfred Nobel, and his father Immanuel. Years before he set up
the Nobel Prize Granting Organization Alfred had discovered
how to set off Ascanio's nitroglycerine with gunpowder and, later,
with mercury fulminate. Now why was Uncle Norman so confi-
dent that such disparate elements would turn solid portions of
Redonda into rubble?

Well, when nitroglycerine is ignited the explosion produces a
large volume of gases and a big increase of pressure as does gunpowder. However, the rapid burning of gunpowder produces

pressures of up to 6,000 atmospheres in a matter of milliseconds while the decomposition of nitroglycerine needs only microseconds and can give rise to pressures up to 275,000 atmospheres. You can keep all this straight if you think in terms of the difference of the effect on the digestion system of a freshly made chili dinner and reheating those wieners and beans you'd forgotten in the cooler for a week.

Despite Nobel's efforts to tame explosions, the problem of premature detonation remained (which was why Uncle Norman shooed everyone away from the boardwalk blasting site). In fact an untimely explosion in 1864 ignited the Nobel dynamite works, killed Alfred's youngest brother and precipitated the stroke that left his father bedridden for the last eight years of his life. Newly motivated by a desire to make a safe explosive, Alfred's experiments led him to discover that three parts nitro absorbed into one part diatomic kieselguhr — silica — made an easily handled dough which, while easier to detonate, had a lower explosive capacity. Alfred dubbed his mixture Dynamite from the Greek *dynamis* — meaning power — and its patent made him a very rich man.

It's not clear who invented the blasting cap that Uncle Norman used to fire the powder that parted sections of Redonda. Nobel himself used gunpowder and mercury fulminate, or fulminate by itself, packed into a copper or aluminum cylinder closed at one end. Much earlier, Samuel Pepys had noted in his Diary how, on a convivial tavern evening, someone demonstrated to him how gold fulmen or aurum fulminans — meaning lightening — when put into a silver spoon and fired, gives a blow like a musket and strikes the spoon downward and not up.

Nobel's creativity was somewhat dampened when his brother Ludwig died and a newspaper accidentally printed Alfred's obituary instead. Nobel unfolded the morning paper to read that not only was he dead but that he'd enriched himself by enabling other people to kill each other efficiently. A shaken Alfred decided to devote himself to rewarding accomplishments that benefited humanity. Recently, campaigners against land mines received the Nobel Peace Prize that Alfred set up with money earned from selling the very explosives used in those land mines.

Subsequent versions of Nobel's original compositions have

been worked out and small delay systems introduced since. As G.I. Brown writes with considerable understatement in *The Big Bang,* his history of explosives, "The precise moment in which an explosion is initiated may be of paramount importance." Various riffs on Nobel's mixture blasted the road and rail beds across Canada and partitioned British Columbia with logging trails. In response to our varied terrain, a variety of esoteric schools of blasting technique evolved. Tree falling provides maximum opportunity for error, as logging accident statistics prove, but dead West Coast Douglas firs are even more dangerous to drop than live ones since the touch of a running chainsaw can cause an avalance of several tons of bark down upon the sawyer. Fallers with a strong sense of self-preservation refused to touch such a tree and Uncle Norman, who ran Joyce Point Logging just north of Teakerne Arm as a sideline, would be called in. Casting aspersions on the size of the fallers' reproductive equipment while ringing the tree with dynamite, he'd string out a long fuse, light it and run. The correct weight of the charge would debark the tree and not Uncle Norman.

In the 1930s Francis Barrow, an indefatigable chronicler of upcoast lore, tied his *Toketie* up at Refuge Cove and "Johnson-ed" up to Teakerne Arm to observe loggers dynamiting apart a vast Davis raft of logs — simply the easiest way. Wood was cheap and plentiful and no one worried that they'd killed everything in the bay.

A view of the boardwalk from the Hopes' porch; Uncle Norman's hardhat on the railing. PHOTO BY JUDITH WILLIAMS, 1987

Ken MacPherson, the epitome of backwoods cool, developed his Laws of Dynamite while blasting roads around Refuge Lagoon. Once a tidal salt lagoon, Grandpa Black dammed it up in the early 1900s perhaps (opinions differ), and it slowly turned itself into a freshwater lake with a salt chaser at high tide. Brackish water lurked in the gloomy depths populated by drowned trees so spooky under toes when one swims out too far. Recently, studies of the lagoon's suitability as salmon habitat in preparation for coho reintroduction, revealed a curious, live, purple jelly that hovers between the salt and freshwater levels.

A prospector and logger, George Black had settled 15-acre Lot 4936 back on the lagoon, where he and Grandma Black produced Ollie, Florence, Jennie, Beatrice, Bill, Bob and Morris. George drained three swampy mead-

ows leading toward Black Lake for his farm and, hard to imagine now, a Union Steamship calling regularly would transport his produce to southern markets and return with cash. The Black children, who rowed down the lagoon and walked through the flume to the Refuge Cove school, grew up surrounded by the tools of the logging trade: dynamite and dynamite caps.

Jan Hanson and Paul Emmons in Refuge Lagoon sending logs to the flume in May 1988. PHOTO BY JUDITH WILLIAMS

"These two, when kept apart, were not dangerous," Doris explained one day. "But the children hearing this and apparently seeing the men bite the caps to set the charges did likewise, and Bob Black had one blow up in his mouth and it was permanently deformed."

She then went on to list all the places that Bob and Nellie Black had lived in during their "logging" lives, because when you finished one logging show you attached a tug to your floating compound and moved to the next. Hence the floating houses.

Once, when Nellie was visiting Doris, she told me that she and Bob had started their married life on a floathouse at the head of Toba Inlet. "We were tied up with Morris and his wife Dorothy near the ice cave. Avalanches crashed down all winter. It was awful," she shivered. "When the stock market crashed no one would buy logs. We divided what was in the cookhouse and abandoned the rest. We went to Cortes and moved into an empty house that had been built by a Mr. Gregory around a enormous stump."

"Later," Doris added, "Norman moved a house down from Joyce Point Logging for Bob and Nellie. He pulled it up over there just past Wicks' — no a bit over to the left because there was another house up there. Nell and Bob were in the middle, and farther along there was a house where the gazebo is, and one in front of that and oh! one more just past the gazebo. Later Nell

and Bob pulled their house up where Barry is building. There
was a well back there too."

As I closed my eyes and tried to fit all these lives together she
said, "There was plenty of privacy. We had a great time. Norman
and I paid the taxes but they had our permission to pull the
houses up. Supplied them all with power as well. 'Course they all
dealt at the store, all except you-know-who around the corner.
Those bastards, up on our land, taxes *we* paid, they ordered their
groceries from Woodward's and had them delivered to our float,
if you please!" She turned down the volume on the radiophone,
looked me in the eye and said, "There were three floathouses off
the ramp in front of the teacher's cottage."

She meant there were more people then, so it was really a
question to me about why the Co-op won't allow our land to be
more extensively populated and why we want the logging con-
trolled. I want the logging flume in our hands and the gate of the
dam lifted so the salmon can go through each fall to spawn. She
saw people needing work, the value of community.

I also see different societies viewing resources differently. The
Toba ice cave that chilled Nellie's heart was, according to Klahoose elder Joe Mitchell, the freezer where they stored salmon
for the winter.

"You know those calm December days," he said? "Well that's
when the people would paddle up Toba to the end and dig out
the frozen fish." And in fact, there were a few days each year of
great calm and sharp sun — sometime between Christmas and
New Year's — and Doris, who had a phenomenal capacity for
cold, would take me out fishing until either dusk or my incipient
hypothermia forced us home.

After we organized the Co-op, old Bill Black continued to
come to Refuge, open up his house, and fish for a couple of
weeks. One spring, his fibreglass cruiser headed out each day for
the bite at the slack. One evening, the boat stayed at Production
Point for some time. Not so unusual in the days when fishing was
good. After a while it became clear that the boat was going
around in tight circles, so Paul Emmons headed out to investigate. He found Bill laid back in his chair with a big spring salmon
in his lap. Neither showed any sign of life.

Paul towed the boat to the docks, but what to do with the

body? It was now dark. No plane would come until morning. There was, of course, the freezer, which was being stockpiled with blocks of ice for the tourist season. Bill and the salmon were laid on the ice and the door shut. Tight. At dawn the radiophone was fired up, the plane called and relatives notified. Occupied with the phone in the back of the store, no one noticed a man and a woman come up from their boat and open the freezer looking for ice. Until the scream. A stiff drink had to be applied to the woman, ruffled nerves soothed, and enough explanation given to avert complaints. The plane came just in time, and Bill and the fish were sent on their way. It was, all agreed, a very fine fish.

It's the way I want to go.

Ah! yes, back to the lagoon dam. When I first teetered up the crumbling catwalk cantilevered out over the creek and flume, the dam resembled an Anthony Caro sculpture left out too long in the rain. A haywire mess of steel beams supported a steel plate connected to an overhead cable and counterweight. With a little luck, you could cut a bunch of logs, bag them within the lagoon's standing boom and, once the tide was super high, hand winch up the steel plate so that the pressure of the lake overwhelmed the force of the tide. In our time Paul Emmons or Jan Hansen would pike-pole a log over the 10-inch step between the two bodies of water. The outflow from the lagoon would send it down the flume and into a bag in front of Dorothy's house in what's known as "the Hole."

On the land where the creek meets the sea, under Paul's house, is a small midden and before the dam and before the Blacks, the indigenous Klahoose people came here to Thushton and gathered reeds from the lagoon. It is said they used the water-filled cavity in the bedrock just behind the dam for cooking. It must have been one hell of a stew because a bout of archeological enthusiasm once prompted me to bail out what I discovered was a three-foot-square hole.

The flume itself is a long, wood-lined chute. At the sea end there is a slight bend into the creek proper, which is really a collection of boulders that a high tide covers.

MacPherson considered Paul's ponytail effete and the name of

Fluming logs from Refuge Lagoon via raised steel-plated dam at high tide, May 1988. PHOTO BY JUDITH WILLIAMS

his boat, *April Love*, an embarrassment, but needing a hand he decided to teach him his logging and dynamite technique.

The glitch in MacPherson's system was that the logs coming out of the flume would take a nose-dive at the bend and bury themselves under the logs chained like guides lengthwise along the side of the creek. If the subsequent timber hit a inconvenient boulder or wedged itself under the first log, observers on the catwalk witnessed a logjam of alarming complexity. The person at the dam had to be stopped from sending along more logs and someone had to descend onto the logs in caulk boots and clear the jam. It was exciting to watch but far too dangerous to be tolerated. If you couldn't unjam the *embarrass* you had to wait until the tide was low and blow it apart with dynamite.

Finally MacPherson had had enough of the pileups, and he decided to kill two birds with one stone: remove some enemy boulders and take Paul on to Dynamite 202. Pat decided to watch.

The tide went down, the creek dwindled to a trickle, and the culprits were charged.

MacPherson said, "Run!"

Paul said, "How far?"

MacPherson said, "How should I know!"

So they ran along the catwalk, around the bluff and along the rickety boardwalk to Ken's generator shed and got behind that.

Now MacPherson had held to The Third Law of Dynamite: Never use 1 stick where 2 will do. The boulders and the noise, confined by the cliffs on either side of the creek, exploded straight out across the boom, missed the ocean completely, and ended up on Dorothy Thomas' beach. The small stuff flew over the generator shed and landed behind Ken's house.

Every so often, more in memory of MacPherson than in strict necessity, Paul blows a few more rocks from the stream, and yet far from being the paved highway one might now expect, logs still do hang up. Pat, now stationed at the catwalk bend, fires a rifle to

signal the log prodder to stop sending logs over the step. The flume is much improved but the catwalk is maintained in a state just decrepit enough to discourage tourists who have been known to walk up the flume, sink the boat and throw the peaveys in the water.

Once when I was drawing at the lake I heard a woman's voice.

"That's not very nice, that sign saying, 'Beware of Leeches.'"

I decided to be honest. "It's not meant to be."

She departed in a huff. On the way home I dropped in at Paul's and asked, "Why do you say there are leeches?"

He sucked on his pipe. "There might be."

Paul Emmons closing up the log bag in the Hole, Refuge Cove, May 1988.

PHOTO BY JUDITH WILLIAMS, 1987

Lately with not a lot of logging sales coming his way and life so safe he's had to take up skydiving just to keep the adrenaline flowing, Paul has decided to clean up some of the wood left sunk in the lake by the careless old coastal logging ways. He came upon a very large boulder on top of what appeared an even larger cedar butt. Collateral damage from some earlier dynamite experiment, and once too worthless to warrant salvaging, it was now a $2,000 log and he resolved to remove the boulder.

Paul loves a problem, smokes it to a solution. As you might now expect he blew the submerged rock off with dynamite, and up came a log eight feet at the butt. There was only one trouble with this log. It was too big to go through the dam and too wide for the flume and would certainly hang up on even what was left of the boulders. However, there's really no problem. How do you split a log? With dynamite of course.

Ammonium nitrate

You might think, at this point, that you can just go out and acquire some of these joy sticks and rearrange the rockery, but only one person per logging show is licensed to buy and transport dynamite. Paul is the only Refugee presently empowered. In his day as the some-time proprietor of Joyce Point Logging, Uncle Norman had his stash. It was lying forgotten in the old store basement until the building caught on fire in 1962. Norman, aware that over the years, moisture had turned it into a volatile mush, made everyone get as far away as possible while he dashed back into the blaze to retrieve the cash box.

The next day he got on a plane to town and bought the present store barge. His last act, when the Co-op bought the land and store, was to raise it up with the last of the Ellingson jacks from his old logging camp and put it on pilings. He then retired to become the guru to our tenderfoot attempts at backwoods life.

But consider this. Although Uncle Norman was licensed to use dynamite, he was 100 miles from Vancouver and even farther from James Island where dynamite was manufactured and there was, and still is, no road or ferry to the Redondas. Norman's problem was how to get the explosives to what he wanted to blow up without preliminary detonation. In the 1940s and 1950s the Union Steamships came to Refuge Cove weekly, and I'm sure they landed dynamite next to sides of beef on the docks. Just who was carrying dynamite seldom surfaced until incident forced dis-

Probing a stump before dyna- miting. PHOTO BY H. MACK LAINE. BC ARCHIVES AND RECORDS SERVICE PHOTO #G-03321

closure. In 1944, newspapers reports of the Canadian Pacific's *Princess Joan,* carrying 400 passengers round Point Grey, ram- ming the *Squid,* revealed the latter was carrying 25 tons of dyna- mite.

One summer, I took passage on a freighter into a series of coastal logging camps, and while I'll obscure the name of the vessel and one stop to protect the guilty, it's fair to say that this ship is a major dynamite transport north of the end of the main- land highway at Lund or, as the sign there reads: "The Beginning of the Highway" — upcoast point of view being entirely related to position.

After boarding, we waited for Seymour Narrows passage at slack tide while a forklift slotted together various untidy masses of boom chains, steel cable, plywood, cedar siding, and a remark- able number of freezers. Passengers leaned over the bridge rail and asked each other what was what. Bundles of lumber were obvious, a wood stove unmistakable and mattresses and box springs identifiable individually but confusing in their consider- able numbers. A reporter covering the voyage pointed at large numbers of shiny black packages tied with fulsome, clear plastic bows.

"What's that, there, those big black piles on pallets? Gifts for the Devil?"

Once underway, day by day and night by night the ship's handsome red-armed crane cleared the deck of cargo. Finally, little was left but the black plastic bundles. As we headed into a maze of islands and channels toward a new logging camp, the reporter, annotating his tiny notebook, whispered, "The black bundles are dynamite; the caps kept somewhere else."

"Under whose bed?" someone asked.

"Oh! Ho!" says Skipper when confronted. "Not like the old days. Much more stable." The ship was full of dynamite. That evening we entered a small inlet where rapids bubbled up on each side of a vast rock. The extremely narrow pass had to be traversed at just the tidal slack because boils caused by a rapidly advancing or retreating tide could shift the hull either onto the great streams of kelp that rippled over the rock or toward shore. I drifted off looking into darkness and came to hearing Skipper say, "Charlie, where are you? It's the barge."

Silence.

"Charlie, are you past Acteon Pass?" He waited, then hung the phone back up on the ceiling. "That was Charlie." "When I asked him where he was he said, 'Fifteen minutes flying time from Port McNeill.' When I said I was the barge he said, 'No one here now', and hung up."

In dense darkness we threaded our way to the pass, which the chart made clear was a pile of rocks, and a man who wasn't there. "Hope Charlie leaves the lights on," Skipper muttered.

The deckhand swung the spotlight to the left, illuminating a tiny white marker on an islet. He swung it right to another marker on a rock.

"What does the depth sounder read," I murmured?

"Don't have one," said the second mate. "Would only scare us."

We inched forward, kelp billowing on either side of the hull, spotlights creating grotesque bearded heads that swung out from the close shore, revolved, became dogs, then clowns in a cedar funhouse. The light flickered stumps and branches into unicorns and rhinos. Moss fluttered away like giant bats. A dinosaur reared its head.

We stopped, enclosed in a dark globe. The skipper checked the chart. "The next pass is even more constricted."

The lights crawled along the shore. I went out on deck, walked to the stern and, as the ship swivelled to starboard, faced a three-storey floathouse lit up like a gin palace. A figure moved past a window. Retreated. The ship snugged up to logs fitted onto a sea-wall of raw fractured stone and lowered its ramp. A man rolled out of the floathouse, beer can in hand, potbelly barely contained in a red T-shirt under a green, padded vest. There was something in his back pocket with a string hanging down. He opened the back of a truck and light fell across the stage.

Charlie?

On deck, the forklift shifted the black pallets of dynamite. Charlie crossed stage left to right front and placed his can of beer in a stone nest with the elaborate care of a practiced drinker. The dynamite was hitched to the red arm and swung up, out, over — the ship listing — and onto the stage. A rangy black shape leapt from the darkness to join Charlie rooting around in the truck for ropes. The shape climbed into a dinosaur of a machine, which erupted to a start and trundled, clanking and snorting to centre stage, reversed, and backed to one side. The cab swung 180 degrees, and the shovel arm screeched toward the piles of dyna-mite.

Charlie tumbled to the footlights, retrieved the beer, took a swig and replaced it with tender care. Using rather dubious ropes, he tied the pallets to the shovel of the excavator. It lifted the bun-dles. The ropes slipped. The black bags slithered from the plastic bows and slapped sharp to the ground. On deck, the forklift shifted the last bundle toward the ship's arm and ripped a bag. Out onto the deck spilt the salmon pink Amex (ammonium nitrate) used for blasting out roads and which is set off by the blasting caps kept, as everyone keeps saying, "elsewhere."

Skipper got a roll of tape. The deckhand pulled open a knife and cut a piece. The roll gummed up, and he threw the knife hard into the deck and ungummed the tape. Skipper retrieved the knife, cut more tape, and they slabbed the bag back together. The load swung off. Skipper dove within the ship, returned and, with spidery limbs extended by the cast shadows of the lights, climbed up the ramp, up the rocks, up the logs and handed manifest and blasting caps to Charlie, who'd retrieved his beer again.

The deckhand sweeping up the stray Amex said its explosive properties were first revealed when a loaded ship of ammonium nitrate fertilizer, anchored in Vancouver harbour, exploded spontaneously and extensively.

Charlie tossed the caps into the truck, rooted around and returned with steel cable, which he wrapped around the restacked dynamite. A man in Stanfields underwear slipped from the floathouse, crossed the stage and jumped up on the flatbed of a truck. In slow motion, the dinosaur lifted and clanked the bundles onto the truck. The ship folded away its arm and swivelled back into nothingness.

Unwinding ourselves from the maze, was like running a film backward. The moss bats, rhinos, unicorn, and rising kelp all happened faster as the retreating spotlight lifted the leering boogie men back into the trees. Light danced fog out through the tight passage.

"Sometimes," the second mate said, "we have sticks of dynamite three inches round and three feet long for the mine at Texada Island. Still onboard are the cases of the 16-inch long sticks for an Interfor camp."

We pulled into a open bay, dropped anchor and awaited the slack tide. Asleep in my flannel sheets I missed the passage through the rapids.

"Oh boy!" the mate breathes out next morning, "That exit was real tense."

Ripple Rock blast, Seymour Narrows, April 5, 1958. BC ARCHIVES AND RECORDS SERVICE PHOTO D-08490

There is no problem a suitable amount of high explosive will not solve

Once the gentleman of the coast actually got blasting powder into their hands, blowing things up became an obsession. One coastal dwelling, after an itinerant existence floating from one Desolation Sound logging camp to another, had been hauled ashore and set on a foundation that created a low basement. New owners, charmed into a restoration project by the house's authenticity, removed old Gilchrist jacks, rusting cable, pike poles and peaveys from the basement and discovered that the previous owner had stored a road builder's stash of dynamite up in the joists for future projects. The renovators fled. The blasting powder was removed by a demolition expert who took his hat off in professional awe at the sheer bulk of what had become a suppurating mass of nitro and filler.

But one can sympathize. The problem with the coastal landscape is that it is uneven, and on Redonda often vertical. If you fall trees in a pocket of reasonably flat land to make a garden, the stones clutched in the residual roots crumple the shovel. Earlier homesteaders would loosen the rock and sand from stump roots, stuff in a few sticks of dynamite, light the fuse and sit on top. Some people like cocaine. Others are boozers, but I'm assured nothing gives you quite the lift as the whump/thump of a blasted stump.

Naturally, things went wrong. Vancouver newspapers reported the occasional death of overreaching stump ranchers, an unsurprising amount of limb and digit loss, and frequent property damage. On Cortes Island, Elmer Ellingson, deepening a hole in bedrock for a new hydro pole tried to cushion the dynamite blast with a redundant sports car body in order to protect the overhead wires. The blast propelled the car body aloft where it took a quarter turn and landed athwart the hydro line shorting out the power of one-half the island.

But there was one place on the coast where a few sticks of dynamite were clearly inadequate for the task of cutting the wilderness down to size. An American senator once proposed that an atom bomb be used to shorten the underwater twin peaks, known as Ripple Rock, which cause enormous boils, whirlpools, standing waves, and strong currents that endangered the shipping route through Seymour Narrows north of Campbell River.

When the Hopes took over the Refuge Cove Store in 1940s, it was the largest commercial enterprise in the area and, according to Cortes' elders, a more convenient place to shop than the now considerably large town of Campbell River on Vancouver Island. A great deal of tug traffic, travelling north and south via Lewis Channel between West Redonda and Cortes Island, gassed up and provisioned in the Cove because of a coastal peculiarity. Vessels travelling north via the Inside Passage, between the mainland and Vancouver Island, had to traverse formidable rapids at either Seymour Narrows or at Stuart and Dent Islands north of the Redondas. Log booms were regularly towed to Refuge and left drifting at the mouth while the tugs steamed into the Cove for fuel and provisions. Many gallons of gas that appeared on towing companies' bills embarked the tugs in the form of beer, socks, bullets or even shotguns. Shocked indeed were the early countercultural Co-opers when asked to continue such a blatant fraud and in the middle of the night as well. Uncle Norman snorted that tugboats wouldn't come unless baksheesh was laid on.

After a refueled tug retrieved its boom, if heading north, it would try to hit the Yuclataw Rapids at either low or high slack when the salt chuck took a break between the ebb and flood tides. Towing tugs were so slow they could not pass the multiple set of rapids before the flood or ebb recreated the overfalls and whirlpools that could break booms apart.

After breaking their journey in Mermaid Bay, still full of signs painted by bored deck hands waiting the next slack, the tugs would head north, skirting the Devil's Hole whirlpool whose downward rotation has been known to suck stray logs under and propel them violently upward. Perhaps you actually have to see a five-foot wall of water going by to grasp the full effect, but I assure you a 30-foot whirlpool opening next to the boat and dropping down several feet does cause you to consider your immediate prospects. A tide book miscalculation once gave me an interesting few minutes as I slide through Surge Narrows rapids on what appeared to be a rapidly moving sheet of water 15 degrees off horizontal.

One could, of course, avoid the lower rapids and traverse what the Spanish called the *Angostura de los Commandantes* between Stuart Island and the mainland. Should you read your tide table incorrectly, I have it on good authority that this channel can present you with a nine-foot standing wave during maximum tide change. One novice yachtsman, navigating the coast by road map, later whimpered to a friend that no one had informed him of the place next to Stuart Island where a boat must descend a watery flight of stairs. That was the Arran Rapids on a good day. The Spanish record of their 1792 passage through these sets of rapids is noted as the worst incident in their entire trip to and from Spain. Despite instructions from the concerned Homalco people of the Village of the Friendly Indians, the ships spun around in such circles that the crew fell about dizzy.

The mariner's other alternative is to hang a left south of Stuart Island, go through Hole in the Wall Rapids and turn right past Owen Bay. Although all these sets of rapids can be safely traversed at slack tide, they are too dangerous for large vessels.

A more appropriate northern access for large ships, and the one chosen by George Vancouver in 1792, is Seymour Narrows, a two-mile-long and half-a-mile-wide gap between Vancouver and Quadra Islands. Unfortunately the centre of the narrows was "encumbered," as an 1886 *Coast Pilot* put it, by the twin fangs of Ripple Rock. You were firmly recommended to enter at, or near, slack to avoid the danger caused by one peak less than nine feet below the surface at low tide.

The constriction of the narrows caused tides to run at 10 to 15 knots over the 3000-feet-high and 350-feet-wide pinnacles. The

Piano move at high tide. Left to right: Norm Gibbons, Ken Ferguson, Sheba the dog, Tracey Lovell, Michael Gibbons, Bobo Fraser, tourist and Mike Lovell, circa 1976.
PHOTO BY JUDITH WILLIAMS

flood tide caused whirlpools 20 to 30 feet across, vicious crosscurrents and a five-foot standing wave at the Quadra shore.

Of course, the degree of turbulence at Ripple Rock depends somewhat on the height of that day's tidal range. The annually produced tide book is a Redonda household essential. City folk don't give much thought to the twice-daily rise and ebb of the sea or that the height of that rise and fall varies in an elaborate monthly cycle. At Refuge, the cry would be, "Let's move that piano (freight, stove, lumber, house or skidder) at high slack."

The Letwiltock people on Quadra Island claim that, in earlier times when the level of the sea was lower, or Ripple Rock taller, young hotshots would paddle up from Cape Mudge and stand on a peak at the periodic June low tides. Unfaithful wives might be abandoned there to the rages of the tidal bore that were, perhaps, a relief from the domestic one.

However, and this really will get around to dynamite after a while, many voyages were accomplished without incident by the native population who had worked out slack water passage without help of a slide rule. However, the terrible force of the passage's currents tended to draw ships towards the rocks, and this forced large vessels to travel on the outside of Vancouver Island, a longer, weather-fraught route. Ripple Rock was a problem. Not, you understand, for the rock. The problem with recalcitrant nature was, as usual, a human one. By 1956, an official casualty list, begun in 1875, totalled more than 100 ships, including the US gunboats *Saranac* and *Wachusett*, assorted fishboats, tugs, and yachts. One hundred and fourteen lives had been lost. After the US Army Corps of Engineers sank a cable ship there they petitioned Ottawa for rock removal, but it took another four decades, and many additions to the list of Ripple Rock's casualties, before anything happened. By the 1940s the Americans wanted to run their battleships up the coast to Alaska without spilling the preprandial rum, and the locals had taken a personal

dislike to the rock since it interfered with running back to the logging camp when the bar closed instead of when the current dic-
tated.

Just before the Hopes opened their Refuge Cove store in 1942,
the federal and provincial governments had seriously begun to
investigate altering Ripple Rock to increase the traffic needed to
boost economic development along the Inside Passage. At one
time it was proposed to complete the Great Pacific Eastern Rail-
way by building a bridge from Quadra to Vancouver Island using
Ripple Rock as bridge pier. Canadian parliamentarians allowed
the job could be done for $300,000, but contractors were skittery.
No one knew how it could be done, and no one sent in bids,
although at Refuge all those loggers, homesteaders and road
builders who regularly blew things up discussed technique at
length while they waited on the dock on mail days.

"Why certainly. Haul in a barge, anchor it to something or
other, drill into the peak, stuff in the nitro and 'Bob's your uncle.'"

In 1943 after demolition experts — professional, amateur and
armchair — worried the problem out, work commenced. The
plan was for a barge to be moored over the top of the rock by 1,100
concrete anchors and steel cables and to commence drilling into
the peaks. However, the violent water snapped the cables as fast
as they were attached.

In 1945 a barge was positioned over the rock with overhead
cables from wooden spars raised on Vancouver Island and on
Maud Island adjacent to Quadra. One hundred and thirty-nine
holes were drilled, and ninety-three blasted. When a boat, bearing eleven workers crossing the narrows to erect more spars, cap-
sized, and nine men drowned, that operation was terminated.
According to fisherman, Bear Scow, the seine boat *Splendor* was
famous for her spectacular Quadra Island launch during which
she rolled completely over. Interestingly, she was a vessel chosen
to transport dynamite to the Ripple Rock blast site.

By 1958, three shifts of hard rock miners from the Dolmage
and Mason Consulting Engineers worked round the clock tun-
neling to Ripple Rock from Maud Island. Mindful of MacPherson's Fourth Law of Dynamite: there is no problem a suitable
amount of high explosive will not solve, the engineers packed
1,400 tons of Nitramex 2H, one of the most powerful blasting

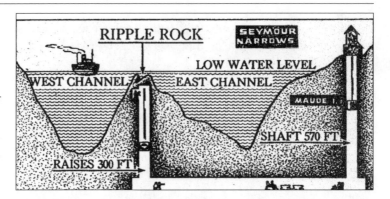

Diagram of
Ripple Rock
tunnel at
Seymour
Narrows.

agents ever developed by DuPont of Canada, into the pinnacle
tunnels. Enough blasting powder, according to Doris, who loved
such statistics, to launch the Empire State Building a mile high.

Pre-blasting plans were elaborate. There was much concern
about the day's weather because low-lying clouds could act as a
blanket that would spread concussion. The tide had to be low
with a strong northerly ebb to dissipate the expected tidal wave
away from development, and that wave was to be measured by
every known device. The small towns of Bloedel and Duncan
Bay were evacuated the day before, and a three-mile -road block
slung around the area. Police patrolled the back roads all night.

The blast, set off from a bunker on Quadra Island by Dr. Dolmage at 9:31 a.m. on April 5, 1958 lofted 740,000 tons of rock and
water 1,000 feet into the air and consumed more dynamite in one
go than had ever been used on the coast. According to Vancouver
Sun reporter Tom Ardies, once the $3,100,000 worth of dynamite
erupted. Mud, rock, water and gas bloomed 400 feet upward like
a strange flower and hung suspended for a moment as its soft pas-
tel shades of light blue and brown entwined. Then it flared up
again, 800 feet high, and spread completely across the 2,500-foot-
wide southern entrance of the Narrows. Riddled by chunks of flying rock, the cloud's loud roar and deep rumble were capped by a
sharp clap like a thunderbolt. Within two seconds, everything was
engulfed in a cloud of white gas that drifted northeast to reveal a
sea boiling as if in a cauldron. An eight-foot-high tidal wave took
five minutes to hit the west and east shores of the Narrows.

The water absorbed most of the impact of the explosion.and,
to the disappointment of onlookers, not a window in the area was
shattered: no shock wave was felt. Uniformed nurses, waiting in
the streets for a possible disaster, climbed into their cars and went

home. The ten ambulances moved off. Mrs. Louise Antifave of Bloedel, who'd wept bitterly that her small house would be blasted flat, was able to move back within the hour. With nothing to complain about, the locals blamed the following heavy rain on the explosion.

Two RCMP boats approached the diminished crags from each end and gauged their new heights to be 45 and 70 feet below water. Considering the new cruise ship traffic, it's not all that much. The subsequent rerouting of ships led to a slow decline in Refuge fuel traffic, and the development of the town of Campbell River on Vancouver Island.

There are certain disadvantages to knowing too much about Ripple Rock. I, unfortunately, have engraved in my brain the report of a fisherman who was heading home south through the gap as he had done many times. He'd sold his fish; the boat was a little light; his family was aboard. Just past the narrows he felt the ship shift abruptly and, sensing it was going to roll, he threw the wife and kid into a dingy and rowed swiftly away. The fish boat turned turtle. Since hearing the story, I go through the Narrows sitting in the inflatable on deck, my life jacket stuffed with protein bars and binoculars, a whistle round my neck, and my fingers crossed.

Two summers ago it was our misfortune to misplace the anchor of our old seine boat the *Adriatic Sea* on the bottom of Shoal Bay due to deferred maintenance on a steel cable that, quite surprisingly, disintegrated into dust during an anchor raise. This anchor lacuna put sufficient crimp in our boating style that we were forced to make wake to Campbell River to pick up our friends, John and Cathy, who were toting another of adequate size up from Victoria. We had to proceed north through the rapids and, at the end of the trip, we had to return our friends to their car in Campbell River, which necessitated yet another passage through the jaws of the maelstrom. That day was notable for a skookum south-running tide. As I turned into Current Passage a flock of logs migrated crossways from shore to shore and a horrified glance at the GPS indicated the *Adriatic Sea* had gone from her usual leisurely 9 knots to 15.

North of the Seymour Narrows rapids we started to back-peddle

The Adriatic Sea
towing boat,
man, and dog
through Seymour
Narrows, cruise
ships in hot
pursuit, August
2001. PHOTO BY
CATHY CAMPBELL

to avoid passage before the slack tide. Then, "Black and white seine boat, I need a tow," crackled on the radio. Ahead was a 35-foot steel-hulled cruiser being towed backwards, and not to much effect, by a nine-foot, aluminum punt rowed by a man and a dog. There was nothing for it but to tie the cruiser astern and proceed as if we knew as much about what we were doing as our classic fish boat suggested.

In the general commotion, we tied the cruiser too close and First Mate John decided to put it on a longer line. It was precisely at the point when the cruiser's skipper disappeared below and the dog appeared to be steering, that the modern navigator's worse nightmare appeared to contest our passage over Ripple Rock. Sliding toward us on the horizon was one of those ten-storey condos disguised as a cruise ship. Behind it towered a second, then a third multi-storey dwelling.

John struggled to keep the cruiser astern rather than parallel while Bobo radioed the cruise ships whose crew, much amused, promised not to run us down. Cathy endeavored to stop John from trying to retie the cruiser and fall overboard in un-rescue-able conditions but, in the interests of marital harmony, seeing a fourth cruise ship on the horizon, gave up and started rooting around in the galley. By this point, I was in a state that would have gladdened the heart of a Zen master. Ever since I had managed to steer or, more accurately, ride the boat unscathed through the log blockade and ingested a medicinal dose of chocolate for shock, I had been beyond care. When Cathy emerged from the galley holding glasses and a bottle of rum, I started to laugh. Clinking glasses, we giggled our way south, falling about with laughter at the sudden appearance of a tug and boom heading north to contest our position over the remains of Ripple Rock. For the marine record, I have to report that the passing of two cruise ships ahead of us smoothed any turbulence into a sheet of glass.

On other occasions I have seen schools of orcas diving and blowing in Seymour Narrows as they feasted on schools of sockeye salmon heading south. But the dynamite that so effectively lowered the Ripple Rock has had a deleterious effect on sockeye heading up the Fraser River to their spawning beds up the inland rivers. The worst disaster occurred when builders blasting the rail route up the Fraser compromised the structure of Hell's Gate Canyon's walls. A photo taken in 1867, before rail construction began in the 1880s, shows Native men fishing at stages built out over the torrent at the narrowest part of the canyon. Women smoked the sockeye which, having lost a good deal of fat content, would keep well all winter.

In 1890 the Canadian National Railroad, rushing construction through this section, dynamited the canyon walls and dumped the resulting debris straight into the canyon constricting the Fraser's flow. A slide occurred in 1912, and on Valentine's Day, 1914 a massive landslide undercut the railbed and a fair-sized chunk of mountain slid into Hell's Gate. This further compressed flow caused such turbulence that water became red with thousands of writhing sockeye stymied in their attempt to reach the spawning beds. The run that would have returned four years later was effectively eliminated and subsequent years' runs seriously minimized. In a misconstrued homeopathic gesture, the Department of Fisheries and Oceans (DFO) planted more dynamite into the slide and blew the debris into a brisk current they mistakenly hoped would sweep it away. Due to the difficulty of the passage, the sockeye run continued to decline, and both commercial and Native fishery were radically diminished.

Building fishways at Hell's Gate Canyon in the 1920s. BC ARCHIVES AND RECORDS SERVICE PHOTO #D-02655

In 1920 the DFO set up a commission to investigate the situation, but because, as you might expect, the appointed group lacked knowledgeable scientists — and certainly no representative of the people who had fished there for centuries — no one could decide what to do. So nothing was done.

A maddening ongoing debate then ensued as to whether overfishing or habitat loss had caused the salmon stock to decline.

The eventual restoration of Hell's Gate Canyon in the 1940s followed a debate on the wisdom of building major dams for hydropower, as was done on the Columbia, or rehabilitating the fish. Oddly enough, we seemed to have voted for salmon, but even more explosives were used to build the vast fishways needed to bypass the rapids caused by the original dynamite. Nonetheless, fish stocks slowly recovered and 1958 was the biggest recorded run since 1913.[1] That dramatic reconstruction on the Fraser was the beginning of the ongoing argument on how to care for our rivers. It led to the interest in the kind of rehabilitation of logging-damaged salmon habitat that has been undertaken in Refuge Lagoon by The Desolation Sound Salmon Enhancement Society.

1. Evenden, Matthew. "Fish Versus Power: Remaking Salmon, Science and Society on the Fraser River, 1900-1960." PhD thesis, University of BC, 2002.

To those who'd mastered the mysteries of dynamite in their previous careers blowing up the province and had frugally toted a stash with them into their new environment, the extensive newspaper coverage of the Ripple Rock detonations was inspirational. The caretaker at Malibu Rapids Camp in the 1950s was a-jack-of-all-trades; logger, fisherman, actor and camp councilors. A Christian Youth camp, Malibu is situated at the tidal rapids that guard the entrance to Princess Louisa Inlet. This narrow granite-walled gorge, cut by an ancient glacier into mountains that rise 5,000 to 8,000 feet above the water's surface, runs northeast off Jervis Inlet. At the far end of the inlet is the exhilarating Chatter Box Falls. One part of the program for visiting Christian youth included an informative boat excursion from the camp to the falls to read a few appropriate verses from a boater's bible, *The Curve of Time*.

The caretaker was a restive soul and found the company a little staid after the raucous Hollywood stars who'd graced the Malibu's previous incarnation. During a few youth trips he'd noted the excellent acoustics of the stone canyon and hit on a way to liven up the excursion by instituting what he called World War III. Taking several go-ahead youths into his confidence, he trained his troops in the intricacies of planting underwater dynamite at strategic spots along the inlet and high in the hill. He sta-

tioned his acolytes onshore with fuses to both sites and, as the boatload of teenagers passed, explosives detonated to spectacular sound and water effect. The Christian youths implored the intercession of their maker to the great satisfaction of the instigator who managed to get in several more eruptive boat trips before the campers informed their parents.

I suspect there's a descendent of the American Senator who proposed the Ripple Rock atomic blast somewhere within the Oregon State Highway Division. When a 45-foot, 8-ton, dead whale washed up on a beach, they sent for the Division's detonation expert on the bureaucratic supposition that highways and whales are both very large things. He proposed blowing up the whale with dynamite. His theory was that, given the Highway Division's ample ammonium nitrate stocks, he could arrange a blast that would atomize the whale into small pieces that could be consumed by seagulls and the creatures of the sea.

Hell's Gate Canyon, 23 miles from Yale, 1867. Native fishing stages and drying racks are on the left. Photograph by Frederick Daley. BC ARCHIVES AND RECORDS SERVICE #A-03874

A fair number of spectators, and a film crew from a local TV news show, turned up to watch. The highway crew moved everyone back up the beach, put half a ton of dynamite next to the whale and set it off. At first the whale carcass disappeared in a cloud of smoke and flame, and happy spectators yelled, "Yay!"

Dynamite box, Refuge Cove.
PHOTO BY JUDITH WILLIAMS

and "Whee!" Suddenly there were new sounds like "Splud!" and a woman's voice shouted, "Here comes pieces of ... MY GOD!" Something smeared the camera lens.

The good humor of the situation suddenly gave way to a stampede for survival as huge chunks of blubber fell from the sky. One sizable piece caved in the roof of a car parked more than a quarter mile a away. Several whale sections the size of condominium units remained stubbornly on the beach. What seagulls still attendant lay prone. Informants tell me that this very sobering videotape is often watched at the end of parties.

A little extra for luck

By the 1980s superannuated loggers Hope and MacPherson had abandoned their battle with the Redonda woods, but once Paul had found a calling exploding things, the masculine Refugees lay awake nights considering how the landscape could be rearranged with the help of dynamite.

Barry has been building his house for 25 years. He claims he is building to last. The last to move into his house? Given the years of labour involved in transporting cinder blocks and, unlike the rest of us, building a cement basement, we were a little surprised that, when he finally had the floor down and the whole thing closed in, he became dissatisfied with the underground configuration and wanted to even out its surface. He invited Paul to study the problem and they decided it would be no trouble at all to peel off a tiresome chunk of the island and thereby widen the Barry's basement.

They drilled a line of holes, stuffed them with powder and padded with blankets and a heavy length of industrial belting that Barry had acquired from somewhere up Lewis Channel. Paul lit the fuse and they run up the hill and hunkered down behind a substantial boulder. They waited for the explosion. They waited some more. Nothing happened. They looked at each other, they looked away. Paul lit his pipe. What was to be done? They give it a while longer. Nothing. Birds twittered, no-see-ums nipped. They couldn't stand it. Paul crawled towards the house, waited, crawled closer, oozed up the wall, and looked into a window. He went inside and, pulling back the padding, found it on fire around the dynamite.

"Well? *Well?* What did he *do?*"

I'm not quite sure, since I wasn't there and the story was told to me by Bobo, who exercises a poetic licence even more liberally than Doris. He might have exaggerated. I do know they subsequently shaved that rock off neat as if a glacier had gone through. So chuffed were they by their success that they studied how to remove a formidable boulder that now interfered with Barry's use of his skidder as a wheelbarrow between his dock and his still-unfinished house. They decided to kill two birds with one stone and remove the boulder by reducing it to fill for the trail. The rock was drilled, MacPherson's Sixth Law of Dynamite ("A little extra for luck") was invoked, and the charge was set.

Oddly enough for such a long-term planner, Barry had neglected to calculate the trajectory of the rock fragments and, instead of being available as fill, rock flying in the direction of the house severed the power line from his generator and punched a fair hole into the cinder block foundation. Barry was pleased; it vindicated his use of the easily repairable cinder blocks. With the boulder gone, the skidder was able to deliver yet another load of material to the expanding building site, even though if he connected his pier to his dock he wouldn't need the skidder at all.

Such calls on Paul's talent were endless. I have heard more dynamite explode here in twenty years than in all my previous thirty-five, and that includes childhood summers spent next to the limestone quarries on Texada Island. There they had the courtesy to blow the whistle before the blasts. However, stumps have the habit of being right where you want to build something, and in a community where there are no roads their removal is a backbreaking job. So . . .

One day, the "jungle phone" that connects Refuge houses jingled. Dixon advised me that he and Paul were about to blow out the stump of a tree at Barnes Bay that had the lack of foresight to have grown where a guesthouse should be. A nest of bottles had been found entwined in its roots. I hustled down to find a glass midden containing coveted cobalt blue Milk of Magnesia bottles, complete with embossed letters. It set me wondering about the really remarkable consumption of this elixir in the old days as evident by the amount found in garbage heaps. An even more remarkable find had been the 12 bottles of after-shave lotion bottles at Dorothy Thomas' place across the bay in the Hole. That was surely a lifetime supply of something dabbed on once a day at most. Did Ed Thomas perhaps swig as well as dab? I accuse no one, but when we first came here there was a unconscionable amount of after-shave lotion in the store considering the roughness of the clientele. Quart bottles of vanilla suggested that the Bachelors baked cakes daily.

According to Uncle Norman, Ed drank more than those 12 bottles of after shave and in fact had a minor brewery of his own, which Norman often felt compelled to row over and inspect. This was a well-known, well-kept secret as are so many things at Refuge. However, just which Ed Thomas Dorothy was married to and Norman drank with, is still debated. Recent intelligence comes from three elderly Englishwomen, now in their eighties. Their father, Walter Edward Thomas came to the West Coast circa 1910 and they and their mother lost contact with him in 1915 after a final letter signed "Your Ever Loving Walt." In 1995 the sisters contacted the Vancouver police who asked a Vancouver *Province* reporter to make public a request for information about their long-lost dad. A photo of Walt was published. Iris Bjerke, 69,

Walter Edward Thomas in a contemporary newspaper illustration.

of Surrey called the paper to say that it was the image of a man who'd arrived at Redonda around 1920 in a canoe, with a bedroll and not much else. Redonda in those days, according to Iris, was the kind of place where people on the run could live and no questions asked. Ha! Maybe not out loud.

As a young girl, Iris lived at Refuge Cove with her very large family, and the migratory Ed Thomas married her great grandmother Emma Jane Nichols who lived in a house on a point. Emma Jane was much older than Ed, who Iris remembers as neat and clean and who looked after her 54-year-old Granny well. "A quiet well spoken man, Ed

could cook and sew as well as a woman, never cursed or swore. I'm 99 percent sure the photo is him. He had reddish-blond hair and a moustache and only one oddity; he never talked about his past or allowed his picture to be taken. He'd walk away."

In 1921 Donna Jackson, later Murphy, arrived for a sojourn as the Refuge Cove teacher for $60 a month. One chapter of Donna's memoirs is titled "The most miserable ten months of my whole life at Refuge Cove." They began when the Union Steamship dropped her on Donley's tipsy dock. A disheveled young man in charge of a dirty, dilapidated gas boat loaded her trunk onboard, chugged across the Cove and tied up to a small boom.

"I climbed off the boat onto the boom and thence up a rocky slope to a very small shake shack. Uncertainty enveloped me. Fred, the gas boat owner, opened the shack door, and I met his aged mother, toothless, drab, stolid, and her husband, good looking, blond, well groomed and definitely much younger than his wife. One quick glance showed the whole colourless room, a stove with a large wood box, a kitchen table and several straight-backed chairs, two shelves on a wall.

Mr. and Mrs. Thomas didn't speak. Fred showed me to my

bedroom and left to get my trunk. I was in a state of shock. The room was quite small. There was a camp cot made up like a bed and two orange crates placed together for a table. The dirt floor was covered with wild goatskins that also served as mattress on the cot. That night, as Ed Thomas serenaded his elderly wife on his violin, I realized we were both the seen and the unseen and I felt strengthened."

Doris Hope and Nellie Black both said the first school was in the Hole, and Donna writes of being able to walk to it from the Thomas's. The filthy one-room school was devoid of supplies. Although Donna enjoyed bringing some order to her pupils' days, she wrote, "I hated my living quarters and I hated the food. The old lady boiled everything, be it fish, ducks, turnips or beans. Many times after dinner I'd rush up the trail and lose it all." At Christmas, she fled home to her family in Victoria, but felt duty-bound to return to her Refuge pupils for the rest of the school year. She discovered an older pupil was suffering "puppy love" and decided to distract him with art projects on copies of the *New York Times* sent by a friend. "Ed Thomas," she wrote, "always devoured the financial pages." When she finally discovered that he had come from Toronto, romantic Donna " suspected he had been a banker and was hiding from the law." Eventually the parents of the vast family of children who made up most of her pupilage finally offered her an escape from the dirt-floored room, and she moved.

Iris Bjerke says that in the 1930s her whole extended family, including Ed and Emma Jane, moved to Burnaby and lived on the banks of the Fraser River. After Emma Jane died in 1942, Ed Thomas packed all of his belongings in a boat he'd built — called the *Devon* — and sailed away. Iris never saw him again. In the late 1950s, Iris' father and uncles were up on Redonda looking for an old friend named Scotty Kerr. Scotty had died, and his widow Dorothy was married to the now elderly Ed Thomas. They reported him active and healthy but a bit absent-minded.

When Doris and Norman arrived at Refuge, Ed Thomas was running the *Devon* as a passenger boat to Homfray Creek and to logging camps up Toba Inlet.

"You know, to the meat man's place," Doris said, "Pop Jackson's camp. Somehow Ed acquired Dorothy, I don't remember how, and one didn't enquire too close in those days, and they lived on the *Devon*. Dorothy wouldn't let the loggers, raving drunk or heavily hung over, inside the cabin. They had to sit on the back deck, rain or shine all the three hours to Toba. Well, that didn't last long, and we bought the thirty-foot *Devon* and used it as our freight boat."

Ed was a pernickety sort of character. He persistently arrived to post a letter on Mondays informing Norman that, as the post office was a government office, it had to be open on weekdays. "And I suppose," Hope would respond, "I should be closed on Sunday when the Union Steamship comes in and everyone expects to get their mail and groceries."

It was the habit of the locals to turn up Saturday and party all night until the Union Steamship turned up at two in the morning. Doris and Norm would meet the boat, get the goods into the store, sort the mail, and open so everyone could boat off home. Naturally, the Hopes wanted to sleep in on Monday.

Doris recalls Ed going to England once in the early 1960s, but he died before the Co-op was formed. Dorothy was then easily 75, as secretive about her age as she was about the strawberry blond curls that peeked out from under a white cotton hat. She wore it pulled down low over the dark glasses she had to wear because of an eye accident somewhere between husbands one and two. She'd escaped from England at the age of 19 and all she valued from her youth was a vast store of music hall songs she liked to sing given the least, or even no, prompting. After we moved Carole Emmons' piano over to Dorothy's, I spent several Christmas afternoons singing with her in her long, dark, added-on-to-and-added-on-to house. (One entered through the dirt-floor shed that had been Ed's boat building workshop and, could it have been Emma Jane's earlier house?) Dorothy would recount how she'd met and married her three husbands and how they'd died. At the end, she'd heave a sigh and say," I never wanted to get married at all."

Dick Thompson, "tall, dark and handsome" as a young man when he met Donna Jackson in 1921, lived alone in Teakerne Arm when we first came. Intending to shock us he once said: "Dorothy wasn't really married to Ed you know."

"He would," Doris said when I repeated Dick's remark. "He was like that."

After Dorothy's death, I began the "dig" of what I called the Dorothy Culture: piecing together Ed's old boat building patterns, broken bowls and artifacts. He'd built the fine boat fisherman Kenny MacLeod rowed, facing forward, from the Hole to the store for mail. I prowled Dorothy's site noting each Pond's Cold Cream jar and Alka Seltzer tube, watching Dorothy's daughter Betty hide the clandestine hair dye bottles that had provided the golden curls. Secrets. Teasing out the puzzle of these lives lived on the coastal fringe led to a fascination with coastal tales of places where you could be, almost, whoever you choose.

Dorothy Thomas. DETAIL OF WATERCOLOUR BY JUDITH WILLIAMS, 1986

Was Walter Edward Thomas same person as Dorothy's quiet, old, boat builder Ed Thomas? A recent missive complains that their handwriting samples differ. The upcoast rumour mill grinds slow but it grinds on and that has something to do with writing these stories down. Once I began to record stories people told me, fragmented narratives like Ed's began to accumulate around me like dust balls collect under a bed. It was just as strange and unstoppable. Identified as scribe, people came long distances to tell me things I didn't always want to know. A well-known upcoast pioneer expanded on the subject of his homesteading mother as a cross-dressing lesbian. A Hilda Gunning walked all the way out from the docks to suggest that her mother's flirtatiousness contributed to Hilda's furious father plunging through iced-over Refuge Lagoon and drowning. Schoolteacher Donna writes in her memoir of going back up on the lagoon to visit "handsome" Jack Staniforth at his sheep ranch. It was that same Jack who the Gunnings had been visiting and who Hilda's freshly widowed mother married rather precipitously, well, according to Hilda who was then three. I'm thinking of installing a sign quoting Truman Capote who, when his informants complained he'd betrayed their confidence, whined, "But, they knew I was a writer."

Sometimes an encounter on the dock will retrieve a story from my own storage system. As I scrubbed the *Adriatic Sea* down from her trip north, Alan Easthope, grandson of the originator of the very Easthope Engines that got Ed and Dorothy's hungover loggers up Toba on the *Devon*, docked directly astern. Alan had built the last of those engines and he and his wife were on the hunt for Easthopes in Desolation Sound. I suggested they head up Homfray Channel to the old Lindbergh farm and see if Ivor's engine lay somewhere in the bowels of the old fish boat in which he'd made his monthly runs to Refuge.

In the early days of the Co-op we always knew Ivor was on his way into the bay by the distant *fu-chug . . . fu-chug . . . fu-chug . . . fu-chug* of his single-stroke Easthope. That famous engine, the first put into a fish boat on this coast, transformed the fishery by speeding transport to the grounds and increasing transport time of catch to cannery.

On his monthly visit Ivor, who bore an awe-inspiring resemblance to the aged Bertrand Russell, would shoulder a case of canned milk down the dock. Shocked tourists berated the store's owners for not relieving this ancient of his burden unaware that to do so would unkindly suggest he was on his last legs. The truth was Ivor looked a good deal more antique than we later learned he was; it was the boat that was on the verge of collapse. Each engine stroke flaked paint and strands of caulking from the boat's planks, across Desolation Sound and around Hope Point to the dock. Ivor would buy a sack of flour, five pounds of sugar, that case of milk, and retrieve mail for Eric Lindbergh and his sister Anna, who, more ancient than himself, he cared for on their farm just south of Forbes Bay.

The pacifist brothers, Eric and Herman Lindbergh, had come from Norway to escape the Great War only to find themselves facing endless explosions by loggers blasting roads inland from Forbes Bay next door. Undaunted, they grubbed out a subsistence farm on a rare flat area below the mainland mountainside facing East Redonda. One brother canned, pickled, salted and

baked the vast amount of food produced by the other who, clad in
shorts year round, leapt from boulder to hillock taming the
wilderness. Apples, peaches, pears, hazelnuts, walnuts, and every
imaginable vegetable rained down on the preserving brother.
They raised chickens but didn't bother with pigs; instead they
shot the bears that raided their orchard. Coastal pioneer Jim
Spilsbury recalls the Lindbergs arriving at Savery Island one hot
summer day in the Easthope-powered fish boat with a skinned
bear carcass lolling across the fore deck. Their plan was to sell the
bear to summer residents as substitute pork since, no matter how
much produce they raised, the brothers needed a little cash for
flour, sugar and lamp oil. Unfortunately, the penetrating aroma
of well-aged bear, combined with the heat had a negative effect
on sales.

   In time, their blind sister Anna came over from the old country, and the energetic family carried on until Herman died. Ivor
moved across Homfray Channel and Anna and Eric, by this time
seriously de-socialized, left supply work to him. Concerned rela-

*The Lindberg
farm, Homfray
Channel*, circa
1987. PHOTO
COURTESY OF
LIZ MAGOR

tives, believing the brother and sister should have better care, moved them to Chicago and took over their financial affairs. But the siblings escaped and made their way back home.

Lawyer Bud Jarvis, hearing of them from Bill and Brenda Finche at Portage Cove, undertook to repair the outboard needed to supplement Ivor's asthmatic Easthope. Bud made the long trip up the channel with the revitalized engine which he was immediately accused of stealing.

When Bud stoutly denied this, Eric cackled, "Did ya bring me any money?"

Money and the Lindbergs was a subject well known to the Refuge storekeepers because any Lindberg cheque was routinely thrown into a shoebox for aging. It was with apprehension the store keepers perused anything brought in by Ivor for cash since it could be five years old and, although there was always some doubt if the stale dated cheques would be honored at the other end, the old folks could not be left without their case of milk. Why Bud was thought to be both stealing the engine and delivering money was unclear.

The siblings died, Ivor soon followed, and the fish boat languished. When we visited the farm in the fall of 1990 we squished around the overgrown orchard through bear-processed pear and apple butters. Over-ripe peaches and plums rolled under foot as in some fragrant Heronimous Bosch paradise. Now that no one was left to shoot them, the bears climbed into fruit trees and gorged so thoroughly on fermented fruit they dozed off and came crashing down when they turned over.

Alan Easthope had come north to find what engines remained after the dissolution of his family's business. One had been rebuilt for the Gulf of Georgia Cannery Museum in Steveston, and I'd found one lying on a pallet at the North Pacific Cannery Museum in Port Edward. I suggested that Ivor's engine and the ribs of the fish boat be bronzed *in situ*.

One stick or three

The Barnes Bay stump that opened the signpost back to the Ed Thomas enigma yielded enough glass to bear out Doris' claim the bay was named after Joe Barnes. His wife and their many children who had arrived and departed in a boat shaped like a big shoe. Their brief 1940s Refuge sojourn seemed too short a time for all those bottles. But before Joe. Before Doris?

The Redondas have been occupied by white adventurers since 1911, and stumps and hollows have released vintage ketchup bottles, genuine orange Orange Crush bottles and a Joe Kapp football glass from the 1960s. Joe, frozen in mid-throw forever, was found under the forest duff when Bobo was cutting trees for our log house. Bushwhacking for a trail one day, I dislodged a tiny white china "Scotty" dog looking at a ladybug on his tail. I originally wrote "tale," and no doubt there is one. Donna Jackson said that she and the school children found a skeleton behind the school and used it for health lessons. What happened? She said nothing more.

Glass is surprisingly durable and identifies lifestyle and length of occupation. On the east side of West Redonda I entered a long shallow cave and stepped back against the smoke-darkened roof frightened by what looked like a bomb protruding from the shell midden. No, maybe not a bomb. Prodded out, it proved to be an iridescent, round-bottomed glass bottle. Rubbings from the embossed Gaelic words indicated it was an itinerant 1870s Irish

pop bottle embedded within a Native midden surrounded by pictographs. Soon after, I was invited on a Klahoose tribal expedition
to their old Toba burial ground. We were guided by Joe Barnes,
who'd been raised there by his English father and Klahoose
mother. We stopped at the cave site so the Joe could decode the
pictographs. Although he was familiar enough with that bay to
caution about underwater hazards, he said no one knew the stories of the Native paintings. In fact, more interested in recent
events, he turned away to ask, "Doris Hope, is she still alive?"

All the way home he related stories of Doris, the store, the
rough customers and how they'd bedeviled her, called her "Horse
Face." That was not how Doris told it. How hard had it really
been for her in the forties? By sheer will she made the Bachelors
remove their caulk boots in the house and clean up before dinner. Periodically she got dressed completely in black, added a
huge hat and boarded a floatplane to Vancouver. Literate and
curious, she made a space for herself by rising late and staying up
until 3:00 a.m. She went fishing by herself. She checked each
new word in the dictionary that sat beside her cigarettes. The day
before she died she corrected my spelling. When television first
became available, the steep walls of the cove made reception
shaky, but Doris, not one to be denied any new access to the
world, would row herself and the television set out to the light
point, climb up on the rocks and patch into the electricity. The
power came from the generator next to her house via a long, long
copper line Norman had dragged through the bush in order to
install the light. Powered up at the point, Doris would crawl back
into the rowboat, wrap herself in a sleeping bag and watch opera.
When a like-minded Refugee began to join her in his rowboat
and sleeping bag, wild rumours circulated about their affair. "We
were in separate boats for heaven's sake!" she would exclaim. "As
if he'd have been the one anyway."

When that light point was named Hope Point in 1999, Doris
led a flotilla there to toast the event with the quart bottle of rye
Bobo had, on her orders, provided. None of us except Doris actually liked rye but as a group, although curiously uncooperative in
Co-op meetings, we have a deep commitment to upholding
minor Refuge Cove traditions: good, bad or mad. That convivial
event revived memories of Norman Hope's memorial service.

Two boats of those concerned motored to the outer end of Centre Island, near the mouth of the Cove, proper, and consigned Norman's ashes to what we believed to be an out-flowing current. Distracted by the ceremonial bottle of rye passed around in his memory, we noticed too late that the ashes were flowing determinedly back to the gas dock, the focus of Norman's life for 45 years. It was deemed only respectful to add the now empty bottle to the ashes and let Norman carry on as he wished.

"The old bastard never would do what he was supposed to," Doris muttered.

What stories Joe, Hope, Doris and Dorothy told me are unembellished fact? Are life's events reworked and polished for a series of audiences until the past is bearable and present laughter obliterates loss, fear, or humiliation? Are stories a form for memory that distances failure or pain? When the Co-opers first came we would work on our log buildings in defiance of Doris and the terrain and tell each other our disaster stories until we were sick with laughter. Norman would say, "You stupid bastards!" Retold, that life seems lived like a novel.

"Of course if you would stop interrupting yourself we can get back the main topic," Doris would interject into such theorizing.

Joe Barnes at the Klahoose burial ground, Toba Inlet, 1986.
PHOTO BY JUDITH WILLIAMS

Doris Hope fishing, circa *1948.*

Well, sure. "Okay."

As I was saying, I took those Milk of Magnesium bottles, vintage pop bottles, assorted eras of ketchup bottles, and the guys took the old beer bottles and blew out the Barnes Bay stump

Empty beer bottles create a kind of vacuum that demands something flow within and be contained. Our neighbour Reinhold totted up his beer intake, compared it with the cash flow from his seasonal burger stand and decided to brew his own. His PhD in chemistry had, of course, led him to a series of yeasty experiments with locally available substances and he'd even taken a scholarly interest in duplicating the spruce beer that Captain George Vancouver's men made at Teakerne Arm in 1792. I often think of the sailors leaving gloomy George on board the *Discovery* scribbling in the log, "Our situation is truly desolate," while his sailors repaired to nearby Castle Lake with a keg of beer for a drink and a soak. Desolation Sound indeed!

Reinhold had decided the only brewable substance we had in abundance on West Redonda was the salal berry. Personally, I don't know what everyone complained about. The wine had aroma, a body you could chew, and the colour of antique stained glass. Its secondary note of burnt skunk was memorable. Once he scored the beer bottles, Reinhold set about with considerable care, an extended perusal of relevant literature and a lot of discussion — well mostly monologues — and began more conventional brewing. The initial batch was good and it was cheap. The more he drank, the more company he had, the more he saved. He couldn't go wrong.

But his house, when he left the Cove in the winter, could and did freeze. One often froze in it even when he was at home, Reinhold not being one to ravage the forest to provide such fripperies as light or heat. Keeping the beer from freezing was a problem.

Keeping the beer at all was also a problem and, although he would not stoop to simply sauce-panning it up out of the brewing tank like another Co-oper who thereby eliminated the tedious need for bottles, new or antique, Reinhold needed to get it into frost-proof storage.

He needed a root cellar. Lights flashed, bells rang: "Two birds with one stone!" You see; he had another problem. In the summer he liked to have his upper balcony dripping with geraniums. He'd built yards of boxes. The balcony looked good from the water. It was Bavarian!

The trouble was that filling the boxes with plants had cost $200 and if he did that each year it, like beer, would ruin him. He could, as I pointed out to him at length, over-winter the plants, but if the house froze so would the plants. He tried a plastic tent on the porch. No luck. The next year he bought more plants but clearly he had a problem.

A root cellar. Certainly! Store the beer and the geraniums. Other people just weren't thinking. He would dig a root cellar. There was only one problem. Backing his house was a huge cliff fronted by a loose tumble of rock. He not so much dug as shifted rocks, sand, and forest duff into new arrangements. Soil, as always on Redonda, was in short supply. When we come across actual dirt we stop and admire it.

Coming to inspect, I noted he'd gotten rather speedily down to what a non-geological speaking friend had called the "deadpan," the essence of Redonda. The cellar was certainly not deep enough and had a unhandy protuberance on the right side. What to do? "*Mein Gott!*" The solution was obvious. A stick or three of dynamite would take off the protuberance, deepen the hole and provide the ever-popular fill.

Paul Emmons smoked over the situation. Since the root cellar had been placed at a convenient 20 feet from the back of the house, Reinhold placed a precautionary sheet of plywood over the glass in the back door and laid a sheet of cardboard over the brand new $2,000 inverter on the porch. The geraniums hung cheerfully from their boxes off the front porch facing the sea.

The workers placed the charges and battened the whole thing down with a conveyer belt.

The jungle phone rang at dawn. It was Reinhold.

"You probably won't hear a thing, but just in case you do, I'm just going set a little charge a bit later. You won't hear a thing, but anyway I thought I'd tell you."

"Fine okay," I said, half asleep. You see some people had gotten a little testy about parts of the island flying past their windows and being woken out of a snooze thinking the world was coming to an end or worse. So he phoned and I got my coffee and crawled back into bed to draw, perhaps to define the word "specism," which people insist I invented.

Ka-Boom! It sounded as if they had blown up Centre Island.

Silence. Ah! Success and peace.

In the afternoon Reinhold phoned and asked would we like to come over for a beer and see the root cellar. Well sure, "Okay." Without TV you tend to look at everything.

Actually, it was pretty surprising. Despite not calculating the effect of blasting rock out of a cliff-backed hole, whose mouth opened toward the house, they managed to displace several considerable boulders onto the porch, blow its roof off, punch holes in the house shingles and wing the conveyor belt up over the two-storey domicile and into the water. The front deck, although not strictly in need of it, was covered with the "fill." The geraniums, untouched, had the smug look of those who, standing next to the avalanche, were not swept away and have resolved, with new purpose, to live a better life.

Reinhold seemed quite pleased. The root cellar remains a large, much-admired hole in the hillside, and future archaeologists will no doubt deduce an aborted mining project but not, surely, brewing and horticulture.

Of course Reinhold knew he could get an audience to stand around and admire the devastation because of the cougar story, excuse me "The Cougar" story. Like the dynamite stories, people have it all wrong. Their reports lack background and context. They're short on archival research. Sure it happened at Reinhold's house. That much is clear, but views vary on whom was the main character or hero. Opinions differ as to who was the victim. I read a get well card and mash note from a admiring female dachshund in Surrey who sent the dog Fritzchen a picture of herself lying on her back in a dress, if you please. It's clear that hussy thought the Fritzchen was the main character.

Magazines presented the story entirely from the point of view of the dog, and Reinhold was heard to mutter that he'd hardly been mentioned. The narrative became less about the cougar than about storytelling. It reminded me of the old game where everyone sits in a circle, and one person whispers a story to his or her neighbour about their high

Cougar.
PHOTO BY JUDITH
WILLIAMS

school prom in Grand Forks, and that person tells the person next to them the story, and so on. You get the picture. At the end, the last person tells the story out loud to the first person. The story, of course, is unrecognizable and comes out exiting a *souk* in Morocco. A story about dynamite turns out to be about Irish exports in the 1870s. People digress so dreadfully. No one ever really listens.

To speak of The Cougar I have to step back and wash in the background. The truth is that that year — 1992 I think — was the year of the cougar. They were everywhere. One turned up in the underground car park of the Empress Hotel in Victoria. The guard lowered the gate. The cougar was tranquilized and shipped north. At Stuart Island, north of the rapids, a big cat tried to jump into a living room through a plate glass window and left the imprint of his face, rather like Christ's on Veronica's veil, on the glass. So I heard.

At Redonda Bay on the north end of West Redonda, a cougar began to hang around annoying the winter caretaker who called a friend on Cortes Island who had a cougar hound. It was, I suppose, a descendent of the packs of hounds kept by government cougar hunters when the-powers-that-be labelled cougars "vermin" and paid bounties for their noses. The hound, far from treeing so the brave hunters could pick it off at their leisure, was itself attacked by the cougar. The cougar, unlike the dog, did not live to tell this story.

That year we noticed the Redonda deer were in short supply; the roses bloomed and their favoured salal berries withered on the vine. Ferns grew full height. Deer trails, so handy for finding a route in the bush, grew over. In the fall, Paul shot a deer and hung it up in his woodshed. He and Pat sat to dinner. Meggie, the golden Labrador retriever, was outside eating hers. Suddenly they heard a distressed wail. Paul ran out to the shed. His flashlight beam illuminated a cougar with Meggie's head in its mouth. Trying to hit the cougar with the flashlight, he fell on it. With a wail, the cat extricated its teeth from the dog, its body from under Paul and fled. The dog, in some disrepair, was shipped to the vet.

Paul was quite right to hit the cougar on the head with the flashlight. In looking up statistics about cougar attacks two things become clear. They seldom stalk adults, and many attacks have been thwarted by fighting back. Cougars have been killed by hammer and rock blows, by shovels, and discouraged by yells and hot water. Two brave children fought off a big cat with the straps of their knapsacks on Vancouver Island, the main cougar habitat in North America. A man there recently slit a puma throat as its fangs gnawed on his skull.

Logger and author Joe Garner reports that in the 1930s the Redonda deer population was so large, a hunter just sat on a stump and waited until a deer the right size came along. Garner found a logging show on the north end of West Redonda for sale because a cougar had crawled under the house and carried off a full-sized Lab. On more than one occasion loggers limbing trees found a cougar strolling toward them from the butt end.

Joe bought the logging show cheap and I recently walked along the old roads he built above Doctor Bay thinking of how he had dealt with the cougars in an unsporting manner by shooting them from the helicopters his brother introduced into the forest industry. That upcoast past is, I'm afraid, a shifting island of attitudes one must circumnavigate with outdated charts. Many old-style logging practices that loom as dangerous shoals today were apparently invisible at the time.

Since the demise of wildcat logging West Redonda is a largely uninhabited island. What the animal population was before explorers and settlers is hard to say, but after Garner eliminated the cougar the deer population boomed. People on Cortes shot their cougars and harrassed wolves to protect domestic animals.

One winter night I woke to hear a howl so wild it defined itself. Wolves! This was more like it. The emigrant Cortes pack was seldom seen but on the bluff above our house and occasionally on the trails, wolf sign could be poked open to reveal deer hair and crushed bone. As I lay in bed on a Sunday morning, the hair rose on the back of my neck at a primal sound. There, over along the ridge, a long howl and an answer, then more and more voices moving along up high and then down into the Hole, past the flume, finally growing fainter up by the lake. We were lost in listening.

Pat phoned: "Did you hear them?"

"Sure did! Yes! Oh yes! We — it was, it is . . ."

"The wolves," she said, were at the door. The dog went mad trying to get out."

The summer before The Cougar a handsome two-point buck we had dubbed Apollo started to hang around at the studio window eating Spanish moss while I drew him. Soon his consorts, Daphne and Daisy, joined him. Daphne was ambitious; she slithered under the garden fence and ate all the strawberry plants. The next time she tried that our cat Molly stood up unexpectedly and the Daphne bounded off. Pussycats are *cats*.

The deer had concluded humans were less dangerous than wolves and proliferating puma. An adult cougar can eat a deer a week, and a female can produce two litters of four kits a year. If only one of those kits survives . . . .

By the next summer the deer were gone. The wolves moved into the cat territory, and a cougar began to eat our pussycats. The truth is animals are always on the look out for a free lunch, and if the free meal resembles their usual lunch they certainly feel free to take a bite.

Bill and Brenda Finche homesteaded across Desolation Sound on the narrow mainland isthmus at Portage Cove, a natural game trail. They usually shot and canned a bear and a couple of deer a year and, at the time of which I speak, a fresh deer carcass hung in the tree conveniently close to the front door. It was an icy winter. Bill was busy digging out a basement, finding arrowheads and bone needles. Distracted with destroying an archeologist's dream, he had skinned but not butchered the deer.

He and Brenda woke up in the night to the sound of chewing. Bill, *sans culottes*, crept to the window and saw a cougar sorting out the choice parts. Outraged he grabbed his gun. "Brenda," he shouted, "at the count of three, open the door. I'll fire. Then shut the door as fast as you can."

The scenario unfolded as planned, and the cougar, transfixed illegally by a flashlight beam, was shot at, and the door shut. A question remained: was the cougar dead? They woke at dawn, cautiously opened the door to find an enormous dead cat on the doorstep.

Things did not always work out that well. Brenda and Bill were artists; hard edge, colour-field painters and determinedly self-sufficient. They had (Archaeologists are weeping, but it's their own damned fault for lolling in offices instead of getting out in the bush.) roto-tilled the isthmus midden from Portage Cove to Wooten Bay. Full of nitrogen and shell, middens are the best gardening soil on the coast. Alkaline in a largely acid world they will grow most anything. Cash strapped, the Finches grew greens for the Refuge Cove Store. On one lettuce run we had to rouse them from bed, feed them coffee, and listen to the night's events before we could depart with the romaine.

In an excess of self-sufficiency, they decided to make corn liquor, and after brewing up a batch, they dumped the mash at the low tide level of their long, shallow mud flat. The little creatures of the sea could nibble away etc.

At this time they shared the isthmus with three pigs, and the porkers, as is the case with most amateur farmers, had gotten out of hand. They liked the pigs; the pigs liked them. They followed like dogs. It was not possible to kill them and not possible to give them to someone else who would.

The pigs got into the spirit of the brewing and ate the mash. Well good, or maybe not so good. The pigs got drunk, staggered around the beach, fell over and passed out on the mud and clamshells. At low tide. The sea began its inevitable rise. Bill and Brenda yelled and shouted. Three large-scale swine snored on. What to do? It was not humane to let them drown.

Well, when dynamite is considered excessive for material removal, one resorts to something called the come-along. This is a patent device that slowly, and usually with endless hang-ups, may move your large object toward your desired goal without the cable snapping and decapitating you. The resourceful Finches rolled the pigs onto flattened cardboard boxes, bound the packages with rope and winched the dormant swine shoreward. The pigs, and there was some bitterness about this, apparently slept through the entire process and contributed not a waved trotter to their rescue.

By the time we arrived, the pigs were working off their hangovers by foraging along the edge of the midden with a gracious-looking Jersey cow on loan from a philosopher upchannel. When we came to visit in the year of "the Cougar," I noticed an odd grave-sized enclosure at the side of the garden.

"Remember the cow?" Bill asked.

"Sure."

"Well that cow, now I don't know, the pigs seem to thrive on foraging along, but that cow died. We dared not eat her lest she had some disease. We couldn't leave her lying around, and she was heavy, so I decided to dig a pit right next to her and I used the Come-Along to winch her into the pit. Contrary as ever, she landed on her back. The pit proved a mite shallow."

I turned and examined the mini-corral's uprights more closely. There was a faint smell.

"So," Finch continued, "I filled in the hole and ran a length of wire round 'er."

We repaired to the house, entering through shelves of canned bear arranged like some non-schedo science experiment. The fading, modernist paintings looked more strangely out of place than ever. I revisited the previous year's acrimonious debate on the merits of Surrealism.

"I'm surprised," I said, " that you don't like Dali's work better."

Jan woke in the night to a feline scream. No Tom appeared the next day. Monica, fond of her cats, walked them on the beach in front of her house with the dog Dusty as outrider. What was always spoken of at Refuge as "the Cougar" was undaunted and took Kitty from under her feet. Jan, building a trail from somewhere to nowhere as a light entertainment, came across suspicious remains: grey fur, black fur and tortoiseshell tufts in what inevitably became known as Cougar Swamp.

People became alarmed. It was hard to say which was more

dangerous: The Cougar or people walking the trails at night armed against attacks.

In April, Reinhold's parents, and his sister Frieda and her husband Bob Home came to visit. Bob and Frieda went fishing. Don Wicks arrived to quaff a few beers with the old folks. The fisher folk arrived home. Fritzchen, the dachshund ran down the dock to greet them, and all returned to the house with an enormous salmon. People, much distracted by the size of the fish, were extending congratulations all around until a sudden yelp from Fritzchen attracted Frieda's attention.

"Something has the dog!"

Sure enough, the cougar had followed them all into the house and had the dog by the head, its weakest spot. Reinhold shut the door.

Bob tried to open the door.

Reinhold shut the door.

Don decided to kick the cougar in the head to dislodge the dog.

Reinhold bent to grab the cougar.

Don kicked Reinhold in the head. As Don tells it, he then grabbed the cougar by the hind end and wrested the dog from its jaws.

Bob opened the door.

Reinhold shut the door as he ran by looking for his gun and some bullets.

The panicked puma repeatedly leapt the sofa causing Mr. Hoge, prone thereon, to alternatively grab it or whack it with his cane. The poor cat, convinced Reinhold's house was unfit for animal habitation, tried to jump out various windows, thereby dislodging the midden Reinhold had built up on the sills over the years. Candles, magazines, ashtrays, rude wax pigs from German friends, feathers, hunting horns, two-quart beer mugs, scientific journals and a five-pound treatise on aerials flew through the air.

Mrs. Hoge backed the cougar into a corner admonishing, "Shoo kitty, shoo!"

The cougar finally threw in the towel, stood dazed in front of the fireplace and Reinhold walked over and shot it. There are pictures. 1) A very small cougar in a very red pool of blood half on/half off a rag rug. (I never liked that rug.) It was a young female with good teeth and tragically worn-down claws; the tag end of a population that had eaten themselves out of house and home. 2) Frieda tending the wounds of a very surprised-looking dachshund.

Reinhold phoned everyone to come and look. Then he remembered, he didn't have a cougar license. Next day the Game Warden was stern on the phone. Cougars were no longer "vermin," no bounty was to be won and he was to deliver the carcass to Campbell River intact."

The Bearskin.
PHOTO BY JUDITH
WILLIAMS

Instead, Reinhold sank it in Lewis Channel.

Because he'd phoned on a radiotelephone, the story spread like wildfire. Newspapers got wind of the event, and people read the tale in Toronto, Paris and Saarbrucken. Someone sent a copy of a German hunting magazine with the story: *"In den Fängen des Pumas,"* as told by the dog of course. "My moment of terror! How I saved my family from certain death . . ." etc.

Mind you, the dog at least could not legitimately be suspected of being drunk throughout although Bobo insists everyone else was. They denied this.

"Sure," Reinhold said," we were having a sociable drink. Drunk? Certainly not!"

"Why then," I asked, "did you shut the door?"

"I thought I would solve our problem."

"What problem?"

"Of people on the trails at night armed."

The cougar, you see, was not the real problem at all.

Bobo was disappointed. He'd been following the career of the puma and felt it should follow its cat appetizers with dog entrees. Mumbling his mantra, "We got dogs here surplus to requirement," he had started affixing little signs of fingers and paws to trees. On the day of the attack, he had affixed a label, steamed off a tin, onto Fritzschen's back that read, "Cat Food."

Bobo had a grievance with dogs dating back to a bear incident a few years previous. While sharing a beer or five with Reinhold, he would suddenly exclaim, "There's yet another God dammed dog, black, there, right on your porch."

"That is no dog," Reinhold said. "That is a bear."

While Reinhold scurried upstairs and down looking for bullets and gun, Bobo strolled out onto the deck and suggested to the bear that its life wasn't worth a plug nickel unless it disappeared.

The bear stepped off the porch but turned to dispute ownership of the honeybee "supers" that he'd been truffling in.

Bobo phoned home. "If you want to see a bear get over here."

By the time I negotiated the trail I saw a young inexperienced bear dead as a doornail. When remonstrated with Reinhold insisted that, "Bees got rights too you know."

I didn't feel too good about any of this and the next day when I went to sketch the skinned bear I felt worse. It looked like a stocky, barrel-chested man hung from his feet like something out of Goya's *Disasters of War*. Invoking the mantra, "You'll never see this again," that has made it possible to live at Refuge Cove, I doggedly drew each muscle and sinew. Reminding myself that, once dead, not a shred of a slain animal should be wasted, I asked for the skinned head and carried this never-to-be-forgotten, staring-eyed thing and its liver, home in a Safeway bag. I had a plan.

I took the head down to the beach in a red net onion bag and hung it, the bag tightly closed, just below the low tide line. The creatures of the sea might, I thought, eat the flesh and certainly, I hoped, the eyes, and I would be left with a clean skull. The liver was delicious.

A month or so later I pulled up the still closed sack. The skull was gone. I checked here. I checked there. I howled with frustration. Nothing!

I had reason to be frustrated. On occasion, my academic colleagues have accused me of being theoretically challenged, but in this case I did have a theory and I thought staring at the bear skull might help me refine it. The year before I had been a little ill and was standing in my studio moodily staring into space. Along the path that leads back to the woods came a black bear.

The bear looked in the window or perhaps only at its reflection in the glass, and I looked at the bear. We were lost together in looking. I shifted slightly. The bear perceived "other" then turned and ran through the woodshed, up a tumbled mass of rock and fallen trees and onto a cliff. I now knew not to try and outrun what is clearly a Olympic athlete in a fur coat.

How long did this all take? All the time there is. Time simply was not. At some point words appeared: "Bear? BEAR!" I ran down the hill to the house: "Bear! Bear! I saw a bear!"

I had gone from a non-verbal experience to concept, had lingered in that moment before one says, "Look at the beautiful flower," that moment when you most are the flower. Seconds later I knew it and said it. There are those who assert we have no experience except through language. They are wrong. My initial experience of the bear was pre-linguistic. That was followed immediately, I agree, by naming. Naturally. That was followed by telling, . which was followed by a realization of the separate sequence of events.

In telling again, in creating the event's likeness, a new thing is present in a new way, and if what I told appeared to be about bears it might also be about telling.

"Only image formed keeps the vision/yet image formed rests in the poem," wrote Heidegger, about, I'm sure, something quite different.

Naturally, when one retells stories created by a group — shared oral history — one gets a great deal more literary advice than one can use, not to mention the odd dart in the forehead. Bobo was never satisfied that I'd portrayed events connected to the bear on Reinhold's porch in context. He suggested I'd neglected the facts and insisted that I straighten out why the bear had been shot. Based on his telling, one can only conclude that the entire sequence of events, like the root cellar explosion, had to do with home brewing.

According to Bobo:

"The black bear on Reinhold's porch was there because the honey hives were there, and the honey hives were there because they'd been sent upcoast in our van by Reinhold's father, Ernie, the guilty party. When I arrived to transport the bees Ernie told me the bees were still out working, and it was too early to close up the hives. He suggested we go into the basement and drink until it got dark. This allowed Ernie time to fill me in about the police who'd been at his door earlier to report complaints about the five large hives he kept in his backyard, which was why he was moving his hives.

"Two practised drinkers faced with an endless supply of anything alcoholic can do severe damage to their motor and cognitive skills in two hours. When Ernie and I emerged from the barrel room into darkness, in the skewed way of drunks since the discovery of fermentation, we were cannily mindful of the task we had before us.

" 'Now the bees are at home,' Ernie said. 'So. We make a jail.' He produced little wooden blocks, some wire mesh and began closing up the hive entrances."

" 'How many bees do we have here?' I inquired.

" 'Fifty thousand each hive. All hives, maybe 300,000.'

" 'Sure hope none of them can get out.'

" 'Fraser, they're in jail now. No escape.'

"We loaded the five hives into the back of the van and Ernie presented me with a water sprayer. 'Spray my boys with this every half hour. Keep them cool. Bad if they get too hot. Okay boys, goodbye.'

"Next day was beautiful. I had the sun at my back all the way to Campbell River. Even so, I was not at my best. I stopped every 25 miles or so to mist down my 300,000 companions and myself as per instructions. I was able to do this from the driver's seat simply by aiming the sprayer to the back of the van. The bees seemed fine. At the parking lineup for the Quadra Island ferry, the sun still at my back and my stomach beginning to settle down, I was dozing meditatively when a little girl tapped at my window.

" 'Excuse me sir,' she said, 'You've got some bees on the inside of your back window and my mom said I should ask you why.'

" 'Well thanks,' I said. 'I'm taking some beehives to Cortes Island for a friend of mine. Guess a couple got out.'

" 'There's more than two bees.'

"I rushed to the back of the van. Because of the position of the hives, and the fact that the van was packed floor to ceiling, I hadn't actually been able to see through to the back since I started. Christ! The back window was black with thousands of bees. A major security breach had occurred. What to do? Obviously opening the back of the van would allow the bees to escape. There would be relatively little in the way of personal danger to myself or BC Ferry Corporation patrons since the bees, being honey bees, only wanted to go to work. However, such a course of action would have had serious consequences for Reinhold's future honey production. The bees would be homeless because their hive would remain in the van. I doubted the Ferry Corporation would allow hives to be left in their parking lot for the season since potential danger would outweigh any benefit as a tourist attraction. No, the bees must remain inside the van where I would also have to be to continue driving. I jumped back in and proceeded to board the ferry."

There appeared to be a particularly Refuge Cove logic at work here. Lock the cougar in the house; lock the bees in the van. But Bobo, well, let's listen to the remainder of his story.

"I'm not fearful of honeybees. I like them. Besides, these bees were in the back of the van facing the sun whereas I was in the front of the van in the shade. At least I was in the shade until the ferry cast off from the dock and set out for Quadra Island and changed its position in relation to the sun. I'd draw a diagram but I think you get the idea. Since the sun was now coming through the windshield, the bees joined me up front. This was not quite so good. The drive across Quadra Island to the ferry to Cortes Island, saw me in a cloud of bees the whole way. In fact, I was forced to use my squeegee to clear a view hole through the thick mat of bees on the windshield in order to see the road. Fortunately no one at the local RCMP detachment was outside the building as I drove gingerly past. Was there a law against this? Was there a Mountie courageous enough to present me with a ticket?

"When the ferry arrived at the Cortes dock I was accorded free and safe (for the ticket taker) passage on to the ferry. No way they'd allow me to roll down the window to pay the fare. 'You can give us the money next time, if there is a next time,' he yelled.

"I endured the 40-minute crossing trapped in my seat, surrounded by curious onlookers. Some who knew me even offended against good manners by pointing and laughing. However, a bluff on my part, in which I threatened to open the door, put the run on them. I did regret my part in reinforcing the prejudice against the inhabitants of Refuge Cove felt by many Cortes Islanders. A man driving a van full of bees was surprising but not that surprising since it was one of 'those people from Refuge Cove.'

"The final leg of our journey involved driving across Cortes Island to Squirrel Cove to collect Reinhold and continue on to the bees' new job pollinating apple trees at Hannes Grosse's Loon Ranch on the south end of Cortes. The trip was fairly uneventful save for the marvelous change of expression on Reinhold's face when he first noticed that all was not quite right with the picture I presented. He cursed both his father and myself in somewhat obscure terms. Since the bees had not offered to share expenses, I thought this ungrateful.

"I confess I got no small pleasure from the fact that Reinhold was stung a number of times on the brief ride across island as we discussed the possibility of mailing some of the angrier bees back to Ernie. At the Loon Ranch we unloaded the four intact hives and investigated the hole in the problem hive that Ernie hadn't seen the previous night. Since its contents were swarming about in the van, it remained only to shovel the bees out in clumps with sheets of cardboard. Bees do tend to stick together and the opera-tion was not unlike shovelling snow. The fact that we were each stung no more than 20 times proves the docile nature of the honey bee, and I'm happy to report that the hive has prospered in its new location."

I'm suspicious of that upbeat conclusion to Bobo's report because Hannes also has a cellar full of homemade libations. I sense a rosy curtain drawn across the affair after they'd celebrated its successful conclusion with even more than their customary zest. When they arrived at Refuge Cove, Bobo and Reinhold were barely able to speak. Asked about the trip Bobo mumbled, "Fine, just fine."

Doris, like Bobo, constructed stories about past events that, in the telling, brought us forward into the present. Putting her hand to her mouth, she'd say, "His mouth was deformed, he bit a dynamite cap as a boy."

Anne would come in from the store with groceries. As Doris put them away, she would continue. "The caps were supposed to be over there," she'd gesture toward the couch, "separate from the dynamite over here," and she'd wave her hand toward the stove. "Supposedly that was safe. It was when they were together they exploded."

The imagery would remind me of the inevitable duality of upcoast life, that the bear liver was one of the finest things I ever ate. Heartbreak liver. Animals and food, trees and shelter, roads and dynamite. I'd say, "Remember the Bearbeque? And, we'd laugh at the photo Denise had of Doris and Norm, in their Halloween rabbit costumes holding long forks of bruin over a fire. She said it looked like one of the paintings I'd done of over-educated counter-cultural attempts at upcoast survival.

The bear they'd tried to barbeque was not young, not tender and, like the cougar, it had seemed determined to join the Co-op. As with other people not officially voted in, he created difficulties. In 1982 Reinhold was in full production of "Dr. Hoge's Honey" and hives dotted the hillsides. One day he found a destroyed hive. That same day, Bonnie MacDonald found her compost pulled apart. Another hive was disturbed and the Honey Meister got huffy. Bear sign appeared along the trail, and parents feared for their children. Bonnie grew nervous about going to the outhouse. The last straw was when the Gibbons had a party and Reinhold, climbing the hill, saw John Everly on his porch doing something to the hives newly stowed there for safety, and yelled at him to hurry up. When he arrived at the Gibbons he found John already socially engaged. Reinhold retraced his steps in time to see the bear carefully dismantling the hives and calmly carrying a "super" down the steps. The bear declined to run away. "Gutt God!"

The next day Hoge built a log blind up a tree. He stashed decoy hives about and climbed up at dusk to await the bear well armed with gun and gunpowder. He took a flask of slivovitz to keep out the cold. When he regained consciousness, it was light. The next night, just at dark, the bear appeared. Reinhold shot and the bear ran off. Bonnie, up the hill hearing the shot, shouted, "Is it okay?"

Our hunter heard, "Did it get away?" and yelled back, "Yes."

Bonnie took the opportunity to take Lucy to the outhouse and the bear passed her going the other way.

It was decided that the next night all available guns would con-
vene near another hive at Monty Hall's. Monty cracked one jug
of homemade blackberry wine, then another. The hunters, one
by one, dropped off to sleep with their guns cradled in their arms.
The bear got that hive too.

The next night Reinhold was too tired to stay up so Norm,
whose house fronted the blind, climbed the tree. The bear
appeared on cue, Norm shot, and the bear disappeared into the
bush. Afraid he now had a very angry bear close by, Norm waited.
Nothing. Finally, cramped and damp, he slithered down, ran
into the house and bolted the door. In the morning he let out the
dog, who sniffed out the dead bear not far from the blind.

What to do? Well, it was photographed from all angles and it
was decided that the hide must be preserved and the meat used.
There was only one problem: because the bear had lain ungutted
through the night, the meat was decidedly gamy and, despite its
recent diet of honey, tough as a boot. Denise, on a self-sufficiency
high, ground most of it into burger and saved the ribs and steaks
for the Halloween party. Everyone got dressed up. Men dressed as
women and women as ducks. Someone came as a salal bush.
They barbecued the bear. It was inedible.

Denise now had shelves full of bear-burger canned a la
Finche. We were invited to dinner. Denise's brother, who was a
vegetarian, his girlfriend, who was a strict vegetarian, and her
child, arrived for the peace and quiet that city dwellers always
assure me Refuge personifies. Money being short, Denise made
lasagna. We sat down to dinner. The lasagna looked wonderful,
and there was lots of it. We dug in. It had a unique flavour.

Bruce's girlfriend got up and removed her child. Denise
looked around the table. "It's the bear," she admitted.

Bruce left the table.

I ate some more. Bobo took a second helping: "Wonderful," he
said, "just wonderful!"

We laughed and laughed. I think there are still perfectly preserved jars of that bear in the shed. The girlfriend became a professional clairvoyant.

The bearskin was sent to someone touted as a taxidermy expert
and returned with a shameful plaid lining and an offensive char-
treuse fringe. His head had been rebuilt over a plastic skull,

which sported a slavering pink tongue. His eyes were beady glass balls. It was insulting. The Gibbons displayed him for a while, but after my bear encounter I carried him home in a green garbage bag and laid him out in front of the stove where he scared the pants off visiting children and dogs. I got interested in him, would prop him up, put hats on, and take photographs. I did a series of paintings. As a model, he was versatile; alternately menacing, silly or sad. He got full of sawdust. I tried to vacuum him and his claws fell off. He went upstairs and we tripped over his head in the night. Bobo kept trying to give him away. I started to polish the story. Paul had warned, "Keep no trophies."

The bearskin of course was no likeness, just a new thing. Perhaps the story is the same. No bear is really in my paintings, dynamite is a joke, and Doris is freed of the past by telling a story that is a paradigm. Not just how she lived, but a form life can take.

Charlie Mould, aged 21, trapping on the Southgate River. COURTESY
HANCOCK HOUSE

Use a lot and see
what happens

Redonda's four-legged residents, if left to their own devices, obviously each have their unique ways of securing dinner, but you might wonder how the bipeds earn theirs. Since the cougars and wolves ate all the deer, no one fancies puma steak, the clam beds are privatized, and the fishing certainly ain't what it once was, the traditional aggressive interface with the fauna of island and sea has dwindled. Today, many Refugees prey on tourists. There is the store, a dock, and Reinhold opened Hoge's Slow Food. A craft shop was erected with an espresso machine in one corner. There is a sailboat charter business, an oyster farm and a seasonal garbage collector.

But even the earliest Redonda adventurers sought to replace killing dinner with a search for the wherewithal to buy their steak, and many scoured the island for valuable minerals. They found 100 inhabitants of a Klahoose summer village at Redonda Bay using a stone-circle fish trap for catching salmon headed up Lillian Russell Creek. They also found Native petroglyphs referring to seafood-gathering on the east side of Ellis Lake. Ignoring these clues as to what they might do for a living, the prospectors persisted in staking mineral claims. In 1892, De Wolf and Monroe of Vancouver received Crown grants for the iron mine Elsie on the north shore of West Redonda. With the help of a good deal of dynamite, 626 tons of magnetic iron were blown loose and shipped to an Oregon smelter. Molybdenum was struck when blasting for a logging road on the slopes of Mount Perritt. In 1920, approximately 8,000 tons of limestone rock were blasted out of Redonda and shipped to Whalen pulp and paper mills.

Although surveyors found a mighty-pretty pink granite in Waddington Channel, no high-paying gold, silver or copper traces were ever found on the Redondas. The early economy was inevitably based on fishing, the Redonda Bay Cannery and logging. Serious prospectors had to look farther afield.

The Redondas are smack in the middle of what Captain George Vancouver named Desolation Sound in the summer of 1792 during what is thought to be the initial exploration of the area. However, earlier Spanish explorers had travelled as far as Alaska. Juan Perez recorded landfall at the Queen Charlotte Islands in 1774 during his sniff round for the mythical Strait of Anian located, according to rumour, at 60-degrees N latitude by a Ferrer Maldonado in 1588. That elusive Northwest passage to the Atlantic; if found and claimed would control trade for the holding country. Exploration of all inlets was therefore essential. Sharing information was not.

When Captain George Vancouver, aboard the *Discovery* met two Spanish goletas, the *Sutil* and the *Mexicana* — captained by Cayetano Valdes and Alcala Galiano — near Point Grey in 1792, he discovered, to his considerable chagrin, that Manuel Quimper, aboard the Princess Real had already mapped almost as far as the north end of Texada Island. Together, the Spanish and British ships anchored first at Kinghorn Island in Desolation Sound in July of 1792, then at Teakerne Arm from where they sent out longboat parties to chart the inlets and channels. Galiano's map of these "Narrow Waters" and the Spanish record of the trip, including the journey around the top of Vancouver Island to meet Vancouver at Nootka, was not published until 1802 due to hostilities with the English.

That both the Spanish and English portrayed the Redondas as one island seems surprisingly unobservant although the English correctly located Mt Addenbroke on East Redonda, because it was used to pinpoint Prideaux Haven, a deserted fortified village opposite on mainland. The *Discovery*'s botanist, Archibald Menzies, reported that the crew suffered a massive flea assault while inspecting Prideaux Haven, which they dubbed "Flea Village." They were forced to strip, tow their clothes behind them back to the *Discovery* and boil their britches before resuming.

"Tabla" engraving from Viage.

Of course, the village's temporarily absent, two-legged inhabitants had their own names for the area based on usage. Klahoose elder Joe Mitchell called Prideaux Haven Qu'k wamen Pap'jus was the rock bluff on the south end of East Redonda that non-Natives call Horace Head. The Klahoose name means "forbidden to jig cod" because, according to Joe, "There's a big mountain goat down there." Ah pokum, "having maggots," was the Spring Salmon river half way up Homfray Channel and possibly the village that Robert Homfray was taken to by Klahoose chief Yay'k-wum who rescued him three times during the difficult Bute Inlet surveying journey for Alfred Waddington in 1861. The mouth of Toba was called Yee'k'wamen, meaning "about to go up Toba," Ye' choosen meant "starting to go up Toba," and Yee'K'wamen meant "those from Toba" i.e., Vancouverites. K'w'ikw tichenam, "having lots of pink salmon" was the name for Salmon or Brem Bay, half-way up Toba. Crabby Captain George, unable to find the Native people who might have straightened him out about all this or to catch a single fish, added the name Desolation Sound to his log.

During these explorations, Valdez found the "tabla" that gave rise to the name Brazo de Tabla that, through some cartographic transmogrification, became Toba Inlet. A territorial marker, burial board or perhaps some form of calendar, the "tabla" was likely made of wood from one of the very large cedar trees growing at Hihaymin, the canoe-making village on the Klite River. The "tabla" was not mentioned to the British.

While Vancouver continued north via Seymour Narrows, the Spanish sailed to the mouth of Bute Inlet. It's possible that earlier Spanish explorations had gone farther than admitted and that Galiano and Valdez had been instructed to check secret charts for previous finds. It is the very paucity of information about earlier Spanish expeditions that fuels rumours of sunken Spanish ships on the coast. Recently, one surfaced in Bute Inlet.

Desolation Sound has a salubrious climate and some of the warmest water on the coast, but 48-mile-long Bute Inlet is given to wild temperature drops. One hundred mile an hour winds sweep down from Mt Waddington, the highest peak on the coast. Norman and Doris always took pains to warn of "The Bute," a strong wind that flattened the Homalco village of Mushkin on Sonora in the 1880s.

After Alfred Waddington started to build the ill-fated road from Bute Inlet to the Chilcotin gold fields in 1862, homesteaders had come to the inlet and pursued Native tales of gold, silver and copper. Forester Michael Blackstock, in *Faces in the Forest*, suggests that faces carved on living trees just north in Phillips Arm mark not just copper deposits but Native smelting sites.[2]

2. Williams, Judith. "The History of Alfred Waddington's Gold Road," in *High Slack*. Vancouver: New Star Books, 1995

In the 1920s a penurious quartet, August Schnarr, Len Parker, Charlie Rasmussen and Charlie Mould logged, trapped and explored in Bute Inlet. Parker and Rasmussen thought they'd struck it rich one exceptionally cold winter when great 16-inch balls of what came to be called "Bute Wax" rolled around Parker's log booms in Bear Bay halfway up the inlet. Believing it to be a form of petroleum, the boys barrelled up the substance, which proved useful for everything from a healing salve to axle grease. When the wax captured the imagination of scientists in the 1950s its lush gold hue suggested it was derived from pine pollen.

Recently, someone left a message on my answering machine with news of a new Bute Wax find, which led to my discovery of gold. I inadvertently lost my source's phone number and, attempting to return her call, cast about for a Bute contact. A Powell River forester, who knew nothing about Bute Wax, new or old, was about to dismiss me as a member of the species of obsessive cranks the Redondas tend to breed, when he recalled seeing a decrepit barge in the Southgate River whose occupants were engaged in mining. At the same time, a lawyer friend was asked to register Bute gold claims, and the discovery of a sunken Spanish galleon.

Claimants said the galleon was part of a fleet of 15 ships that had been heading south loaded with gold. Detecting, and just how is not clear, the English heading north, the Spaniards turned back, loaded all the gold on one ship and scuttled it in Bute Inlet at the base of the very mountain they had mined. Hmm?

An unadventurously skeptical person might dismiss this story. However, I had seen a photo taken in the 1920s of Charlie Mould holding a string of dead marten. I had also seen Charlie Rasmussen's photograph of August Schnarr in which he is holding a string of marten, taken by . Same string? Same story? According to Rasmussen's diary, he and Schnarr were up the Homathko River trying to turn fur into cash.

Earlier, Rasmussen's daughter Sylvia had shown me a sample of Bute Wax. Seduced by the sight of her liquid gold, I tracked down chemist Judd Nevenzel, who had analyzed the stuff in 1972 and pronounced it a lipid derived from the body of copepods, a form of minute crustacean. This was a disappointment to Sylvia and her brother, who refused access to the family map locating the smoldering stump supposed by Rasmussen and Len Parker to be the land source of the "petroleum" that was to be their financial savior. However, Sylvia did divulge that Parker, like most old-time settlers, had a series of mining claims that were turned into a tidy sum by his widow; not the first English-professor wife who'd published his poetry, but the younger, second wife who favored Bute Wax as a moisturizer.

Now you might think moisturizer and gold have nothing in common, although Estee Lauder did pretty well, but it seems likely that Parker's claims are the ones now registered as "Slumach's Mines" by one J. Mould. Now, if you're still with me, I'll explain that Charlie Mould had ignored the Bute Wax oil rush and, after the marten market crashed with the stock market, he concentrated on pursuing tales of a Salish man named Slumach who brought gold nuggets to New Westminster in the 1880s to finance months of wine, women and drunken muttering. When the gold ran out, Slumach would acquire a teenage girl and head up valley to, it was assumed, Pitt Lake. Channelling Slumach, Charlie Mould divined that he actually kept going all the way to the Southgate Peak at the head of Bute Inlet.

Now that's a fair hike and no doubt taxed his calves as much as the story taxes my credibility. However, due to the maddening tenacity of the story through time, and the Homalco people's claim of a well-used trail from Nemiah in the Chilcotin to the

Southgate River, I have set geographical rigors temporarily aside to show how Charlie Mould's search for Slumach's mine led to my discovery of gold.

When Charlie's son Jack was 16 he was taken up to the Southgate area to see a wooden door carved with heads wearing what the elder Mould had decided were Spanish helmets. Jack found a "bucket" lined with hide and "strange" markings carved into a tree.[3]

3. See M. Blackstock, *Faces in the Forest* (Montreal: McGill-Queen's UP, 2001) for explication of Native syllabics carved in a tree marking death, territory or incident.

A cave, inevitably called the Spanish cave, had been hewn to enlarge its size. Curiously, for that damp climate, the cave was still framed by wood. Just why all this was attributed to the Spanish when homesteaders, whiskey traders, road surveyors, the Royal Engineers laying out the town of Waddington, and the local Homalco people had been all over the territory is odd, but the "door" is intriguing.

The only find the Spanish illustrated in the journal of their travels through this area is the Toba "tabla" that was drawn on the spot by Jose Cardero. Mysteriously, the 8- by-4-foot piece of wood was carved with a series of circular forms: five quadrupeds and a skeletal, central figure. A skeletal figure us often used to represent a Native shaman, and the quadrupeds are identifiable as the local Klahoose people's sign: a mountain goat.

Since archeological site reports identify a number of cave burials containing human remains on the Southgate side of the inlet and bodies were commonly buried in incised boxes, and wooden burial boards were sometimes erected, any wooden object found in or near a cave might well be of Native origin.

Just how the "door" had been constructed and carved would be useful to know. However, and this is to be expected in mythic stories, all the evidence was buried, according to Jack, in the construction of the logging road up river. Interestingly, that road is about to be examined by archeologists and the Homalco for remains of an old Native trail.

In 1934, Charlie Mould returned in Bute to do further prospecting. He claimed he was tracked by a Native man for a gold nugget he'd acquired from another indigenous person and that he killed the tracker. Son Jack grew up to be an equally heavy dude who, according to his authorized biography *Curse of Gold*, spent his life alternately prospecting for gold in Bute or being unjustly incarcerated.[4] It's said that Jack and a friend were sitting in a Vancouver beer parlor when a man entered and shot the friend.

"Jack," said the friend, "I think I'm dying."

Jack looked his friend over, said, "I believe you are," tipped back the end of his beer, and left.

Jack attributed his minimal gold return and a series of accidents to the curse, "Nika memloose, mine memloose" that Slumach set on the mine as he was hanged for murder in 1890.[5] Eventually Jack discovered a letter written by one John Jackson, dated May 28, 1924, which gave the location of a mine. The indicators were a cairn surmounted by an engraved, tent-shaped rock that faced a creek that bubbled in places over bedrock bright yellow with gold, and then disappeared. These indicators could be found by lining up three mountain peaks.

After a helicopter search, Jack found the indicators but was unable to find the mine. How could he quit? Finally he hired a combination dowser/blasting expert to dowse for and crack open potentially gold-bearing quartz veins using the dynamite which is as essential to mining as Nobel had hoped. Now Jack adhered to MacPherson's Final Law of Dynamite: use a lot and see what happens. His crew hauled vast amounts of blasting powder up Southgate Peak to their camp. It's Jack claim that as soon as his back was turned lightening struck his arsenal and blew the camp apart. Jack was thrilled.

"Gold is the perfect conductor," he exclaimed, "I'm closing in on the mine!"

The rumour that drifted down channel to my Redonda scriptorium was that, in the process of putting in a landing stage for renewed mining, a third generation of prospectors had found a Spanish galleon full of gold sunk in Waddington Harbour silt. You bet. As an underwater archeologist said, "Well, we get a couple of those a year." He also said that, if buried within the abundant silt that flows out of both the Southgate and Homathko Rivers, a wooden ship would last indefinitely.

That reminded me of the stone pier constructed at the mouth of the Homathko River. If you look at a marine chart it's clear that

4. Hawkins, Elizabeth. *Jack Mould and the Curse of Gold*. Surrey, BC: Hancock House, 1993.

5. Slumach's body was claimed by his Katzie nephew Simon Pierre and buried within the old jail. Simon Pierre's Shamanic beliefs were the subject of Diamond Jenness' *The Faith of a Coast Salish Indian*. B.C. Provincial Museum, Victoria, 1955.

the centre of Waddington Harbor is very deep indeed and kept that way by the combination of the river flow and current. One day, DFO divers went down into the hole bounded by the silt estuary and concluded that its whole edge was unstable. They dared not touch any of the walls for fear of a slump that might bury them.

Some time later, and, as usual, with no consultation between governmental departments, the Department of Environment ordered Scar Creek Logging, who dumped and boomed logs in Cumsaw Creek, to switch to deep-water moorage, which would have less estuarial impact.

Scar Creek blasted a major road to the harbour and hired a road building crew to blast out more rock and fill, which was dumped on the silt shelf to form a pier into deep water. After several exhausting months, the crew completed the structure with a graceful truck turnaround at the end of the pier. They left the D-8 Cat where it had finished refining the surface of the turnaround and flew out for ten days' R&R.

Fisheries inspector Jim Nelson was going up inlet for the annual herring count and, informed of the wonderful addition, kept his eye on the port shore for the new pier. He glassed the rock shelf with his binoculars and checked his chart. Nothing. Where a new 20-foot pier and turnaround should have been was the same old shore, albeit newly decorated with fragments of black barrier mesh festooning the trees.

The silt bank, a rather weak link in the whole plan, had been pushed beyond its limit and it had slumped down into the hole bearing the D-8 Cat, boulders, lining web and fill. The tidal wave created by the sudden slump had re-sorted the material and dumped it back 20 feet above the tide line. The D-8 Cat, not a notoriously amphibious vehicle, was never seen again. With the silt shelf gone, the ecology experts, throwing habitat protection to the winds, allowed the blasting of a football-pitch-sized ledge onshore so that logs could be dumped straight from shore into what they presumed was now deep water.

I have news for the bureaucrats. Some years ago, exiting the river near the log dump and assuming the water was deep where the booms were moored, we hit something with the prop. A D-8 Cat on a silt shelf?

Nothing small ever happens at Bute. The wind blows the hardest, the temperature drops the quickest and farthest. Sixteen-inch balls of wax roll around on icy days. The place is off the Guinness record book scale. Ripping down in the spring melt from its vast drainage, the Homathko River rearranges its banks each year and dumps large amounts of silt into the inlet. The gentler Southgate does its part. A wooden ship, sunk in the 1700s and slumped farther down over time, might be preserved. It could be recoverable. Maybe. The claims lawyer, promised a confirmational underwater video, awaits it still.

How stories like this grow, how nuggets of fact are found, rearranged, transposed and beaten into, well, I could go on, but the story is pure gold. Combine a Native curse, Volcanic Brown and his buried jars of nuggets and his missing body — a story Ivor would tell of the hills behind Forbes Bay — add a sunken galleon and bones scattered on moss, and what we have is an archetype. An "Ur" story: a boy's adventure and the missing girls' tragedy. Remember the teenage girls? They never returned from their journeys with Slumach.

Many ounces of gold were taken out of Loughborough Inlet in the 1930s. Is there gold in Bute? Could be. That the Spanish were there in the 1700s, mining and smelting with slaves (did I mention the slaves? — naturally there were slaves), well, perhaps. Evidence of Spanish presence is reported all over the coast. Much of it proves false. But I am instructed recently that Sir Francis Drake, who raided Spanish gold-bearing ships, may have been in this territory in the 16th century.

And consider the story of Huei Shan, a Chinese monk who, according to the writings of one Ma-Twan-lin, travelled from China to the land of Fusang where he dispersed Buddhist precepts to people who derived their clothing, boats and building material from a large tree in — whoa! 499 AD? One family in the Queen Charlotte Islands still claim him as an ancestral visitor.

Don't you want to believe?

The constant is the tall tale, containing just enough verifiable elements to undermine disbelief, and how it moves through groups, is transformed and persists. Charlie Mould's gold is his story.

Doris and Norman Hope's house, Refuge Cove. The blasted rock, boardwalk and rose are to the left. PHOTO BY JUDITH WILLIAMS

Epilogue

Like the story of Ed Thomas, my discovery of gold is a work-in-progress, but some minor mysteries have unfolded at Refuge to what is considered a satisfactory upcoast conclusion. Dixon, unlike Barry, finally attached a float to his pier. When Denise came for a visit, we decided on a ceremonial swim to Barnes Bay. As we rounded the point and paddled toward the dock, we were greeted by the figures of the Dixons, *père et fils*, trimming off the cross planks they'd attached to the enormously long cedar trees installed as stringers for the ramp. Waving the saw in greeting, Pere Dixon revealed that was carrying out his carpentry *sans culottes*. Still clutching the chainsaw, he was at pains to demonstrate the trampoline effect created by the elastic stringers. He leapt up and down like the lead dancer in an excessively Canadian staging of *L'après-midi d'un faune*.

Rick Carter returned for a two-week vacation that same summer and unravelled the mystery of the rose. As usual, parts of Doris' house were falling into the sea, and the accoutrements of what passes for civilized living at the Cove were refusing to do their stuff. The toilet Rick had spent his previous visit installing had become cranky. The day he undertook its renovation we were down the new boardwalk at the "Middle Cabin" drinking hello to Mark Ruwedel's parents.

A neighbour arrived to announce that Rick had plumbed the depths of the problem and discovered that "where the drainpipe

attached to the toilet" no longer existed. Whatever entered from
above descended directly below, and below was a 16-inch tumble
of irregular boulders. How long the situation had existed was
unclear and uninviting to investigate.

Solutions were thoroughly discussed by the party on the boardwalk as they sipped their wine and munched *hors d'oeuvres*.

After some time, I suggested that the visitors might not be as
deeply interested as the locals. I had noticed their drinks suspended at half-mast and their canapés stall untouched in front of
their appalled lips.

Always eager to bridge a social embarrassment by flogging a little local history to a new audience, I explained that Doris' house
was one of the original floathouses from the 1900s. The steep pile
of rocks on which it perched could never have occurred to you as
the site on which you could establish a foothold, let alone a
household. Uncle Norman, however, was made of sterner stuff.
After extensive foundation work with dynamite, the floathouse
was hauled up, jacked up, and pilings applied underneath. The
bottom was closed in and became a workshop containing the jungle phone system that connects the Co-op houses and businesses.

The previous winter I had slept in the back bedroom where
the glorious New Dawn rose taps on the window to come in.
When I'd climbed into bed it occurred to me that I was cantilevered out over a considerable drop by questionable cedar pilings and dubious joists. Perhaps the rose held things up. The
point was that the building was crumbling, and getting on or
under any section to affect repairs unappealing. We have to
weigh the people who are allowed to get on the roof for fear it will
collapse. Those sufficiently light tend to spread themselves out as
if on quicksand.

"So?"

After the sundowners, we strolled back along the boardwalk. As
I passed the corner Uncle Norman had smoothed out by lofting a
portion of Redonda skyward, I found myself staring straight at the
rose rambling over the front of the house. Below it was exposed
the broken remains of an old ceramic sewer pipe packed solid
with whatever packs drains solid. I looked up at the rose's glorious
blooms. It was a tribute to its species. Masses of perfect pink blos-
soms had gladdened our hearts each year. Its leaves shone; the

canes were thick and vigorous; it had never known blight or black spot. And this was odd because it appeared to grow right in the stream that descends from the hillside above the house and runs down under it to the sea. Lightly watered in summer, it stands in a winter torrent.

Now gardening books are insultingly dismissive of gardening in the deep forest or in areas devoid of real soil such as we do, but I had believed them when they said that roses did not like wet feet.I studied the rose. It was flourishing. I gingerly studied the old pipe. It was plugged. I grew suspicious. Was the rose the culprit? Had it invaded the six-inch pipe, grown uphill protected from the creek, plugged the drainage and built up enough weight to pull the pipe apart? Rick had come in the nick of time, for the rose, I began to see, was ambitiously heading up into the toilet to invade the house as had the ivy that decoratively circled the living room.

Old calender published by the Refuge Cove Store in the 1950s. The photo is of Norman Hope's camp at Joyce Point Logging, Lewis Channel. The cabin is now our "Middle Cabin." PHOTO BY JUDITH WILLIAMS

We decided to drop in and console Doris her at this misfortune. I worried she might find her often-mentioned fastidiousness compromised. *Au contraire*! She was eager to put her foot up, order drinks and evaluate the situation at length.

The solution was not so easy. Once discovered, the unsanitary circumstance had to be remedied. Someone had to go under the bathroom and connect a new pipe. The short space was dangerous and lacked aesthetic appeal. What to do? Bobo cheerfully pointed out that at 6'6" he did not fit. Rick is almost as tall and lithesome Alan, who is allowed on the roof, had thoughtlessly sailed off to the Charlottes. What we did have on hand was an excess of teenage boys. Bradley sensibly declined, but young Sandy crawled underneath, connected a new plastic pipe and

Doris.

earned every Loonie of his considerable reward. The rose was all set. In typical upcoast fashion, the old pipe was left where Norman had placed it: in the tumble of rock he'd dynamited roughly level, and by the time the New Dawn works its way up through all that fertilizer, the new plastic pipe will have become brittle. The rose can invade that and grow back down the hill. The remaining mystery was, of course, why no one smelled a thing.

It is often maintained by their followers that the bodies of saints are incorruptible, and that they are proceeded by or exude an odour of sanctity. While comparisons to Saints Theresa or Francis might not be the first, or even last, thought that springs to mind when one encountered the lady of that house, nonetheless, I offer the modest proposal that we had harboured more than what we referred to as the Duchess of Refuge Cove.

Here was a woman who, in her sixties, for no good reason, adopted a large group of overeducated, underskilled egomaniacs and made her home, and all she knew, available to them. Doris showed us where and how to fish, instructed us on whistling down eagles to impress our friends, how to make fish and chips for 30, and to lose gracefully to her at poker. We were allowed to invite stray, awesomely coifed "back-to-the-landers" to her place for massive Christmas dinners. She entertained all the tourists so we didn't have to and lectured us on how to avoid interfering in other people's affairs while telling us the juicy details of their lives. She answered my every query about the past with stories that have become my literary models. On top of this, well, I really don't know how to put this delicately, but, this is a woman whose, well you know, smelled like a rose.

Acknowledgments

Dynamite Stories is drawn from the oral history of Desolation Sound. Thanks to all Refugees past and present for their narratives. Norm and Denise Gibbons, Reinhold and his family, John Dixon, Bonnie MacDonald and Bobo Fraser laughed at their portraits instead of suing and, with current Refuge Cove Store owners Colin Robertson, Norm Dowler and Adrienne Janner encouraged me to publish the stories. Doris Hope, as I trust I've made clear, not only contributed many of the stories but her verbal delivery, vocabulary and narrative structure were pinched to create the style in which the stories were written. Anyone who knows him will detect elements of my husband Bobo Fraser's storytelling mode throughout.

I am grateful to Dorothy Thomas' daughter Betty Yerex and to Iris Bjerke for the Ed Thomas material. Any further information about this person, or persons, will be integrated into the Redonda Islands Archive.

TRANSMONTANUS is edited by Terry Glavin. Editorial correspondence should be sent to Transmontanus, PO Box C25, Fernhill Road, Mayne Island, BC VON 2J0.

New Star Books Ltd.
107 - 3477 Commercial Street
Vancouver, BC
V5N 4E8
www.NewStarBooks.com
info@NewStarBooks.com

Edited for press by Melva McLean
Cover by Rayola Graphic Design
Map by Eric Leinberger
Typesetting by New Star Books
Printed & bound in Canada by AGMV Marquis
First printing June 2003

Publication of this work is made possible by grants from the Canada Council, the British Columbia Arts Council, and the Department of Canadian Heritage Book Publishing Industry Development Program.

NATIONAL LIBRARY OF CANADA CATALOGUING IN PUBLICATION DATA

Williams, Judy, 1940–
   Dynamite stories / Judith Williams

   (Transmontanus 1200-3336 11)
   Includes bibliographical references.
   ISBN 0-921586-95-7
   1. Desolation Sound (B.C.) — History. 2. Desolation Sound (B.C.) — Biography. I. Title. II. Series.
   FC3845.D47W54 2003   971.1'31   C2003-910822-8
   F1089.D47W54 2003